Fundamentals of Power Electronics

FIRST EDITION

BY PRASUN BARUA

ABOUT

Welcome to Fundamentals of Power Electronics! This is a nonfiction science book which contains various topics on fundamentals of power electronics. Power electronics is a branch of electronics concerned with the use of electronic devices and associated components in the conversion, regulation, and conditioning of electric power. Converters control the key aspects of electrical power like voltage, current, frequency, and the basic form of AC or DC. Power electronics is able to convert and control of electric power. Power electronics systems are widely used for various applications which contribute in global industrial and social activity. Power electronics have significant impact on daily lives of people, from cellphones to pacemakers, utilities to autos. Power electronics also allow for the variance of electric motor drive speeds, reducing the amount of energy consumed by making processes more efficient. By converting the Direct Current energy produced by solar panels into AC used in the commercial electrical grid, power electronics allow solar energy to be used. Despite variable wind conditions, wind energy must also be processed and sent into a grid at a steady frequency. Other forms of alternative energy, like thermal, hydro, and nuclear, rely on the capabilities of power electronics to supply electricity efficiently. Applications of power electronics include temperature and lighting control, solar power and renewable energy, medical applications, computer networks and data centers, military, electric power networks and transportation. This book covers various topics on Thyristor, Circuit of Thyristor, Triac, IGBT, Diac, UJT, SMPS, Transient Suppression Devices, Solid State Relay, Single and Three Phase Rectification. This is the first edition of the book. Thanks for reading the book.

TABLE OF CONTENTS

CHAPTER-1: SILICON CONTROLLED RECTIFIER (SCR) - THYRISTOR

The Silicon Controlled Rectifier, or SCR, or simply Thyristor as it is more often known, is comparable to the transistor in many aspects. The "silicon" component of its name refers to the fact that it is a multi-layer semiconductor device.

It requires a gate signal to turn it "ON," as indicated by the "controlled" part of the name, and once "ON," it operates as a rectifying diode, as indicated by the "rectifier" portion of the name. In actuality, the thyristor's circuit symbol suggests that it functions similarly to a regulated rectifying diode.

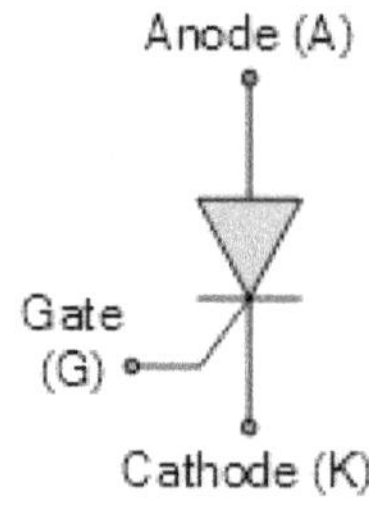

In contrast to the junction diode, which is a two layer (P-N) semiconductor device, and the commonly used bipolar transistor, which is a three layer (P-N-P, or N-P-N) switching device, the Thyristor is a four layer (P-N-P-N) semiconductor device that contains three PN junctions in series, and is represented by the symbol shown. Similar to diode, thyristor is a unidirectional device, meaning it will only conduct current in one way. However, unlike the diode, the thyristor can be configured to behave as an open-circuit switch or a rectifying diode depending on how the gate is triggered.

To put it another way, thyristors can only be used in switching mode and not for amplification. The silicon controlled rectifier (SCR), along with Triacs (Triode ACs), Diacs (Diode ACs), and UJTs (Unijunction Transistors), is one of several power semiconductor devices capable of acting as very fast solid state AC switches for controlling huge AC voltages and currents. As a result, for the Electronics student, these solid state devices become quite useful for regulating AC motors, lighting, and phase control.

The thyristor is a three-terminal device with labels such as "Anode," "Cathode," and "Gate" that consists of three PN junctions that may be switched "ON" and "OFF" at an incredibly fast pace, or "ON" for varying lengths of time between half cycles to give a specific quantity of power to a load. The best way to understand how the thyristor works is to imagine it as a pair of complementary regenerative switches made up of two transistors connected back-to-back as shown.

The Analogy of Thyristors and Transistors

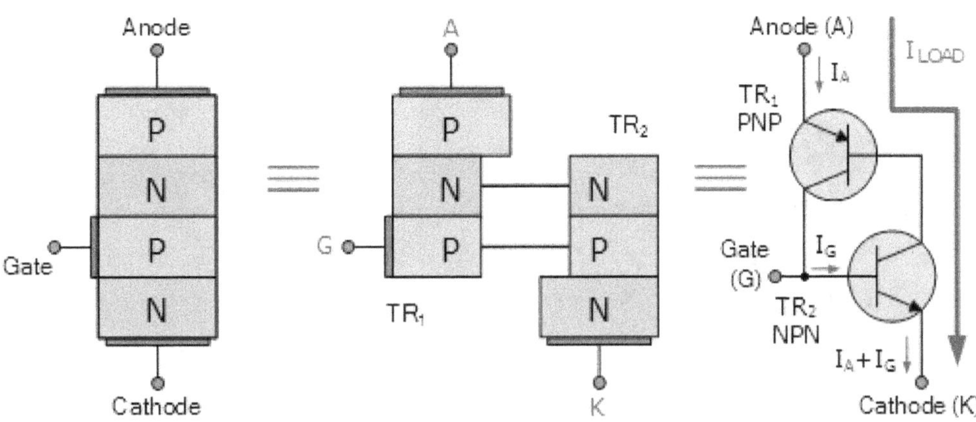

The collector current of the NPN transistor TR2 feeds straight into the base of the PNP transistor TR1, whereas the collector current of TR1 feeds into the base of TR2. Because each transistor obtains its base-emitter current from the other's collector-emitter current, these two interconnected transistors rely on each other for conduction. Even if an Anode-to-Cathode voltage is present, nothing can happen until one of the transistors is given some base current.

When the Anode terminal of a thyristor is negative in relation to the Cathode, the center N-P junction is forward biased, but the two outer P-N junctions are reversed biased, and the thyristor functions like a regular diode. As a result, a thyristor inhibits the flow of reverse current until the breakdown voltage point of the two outer junctions is exceeded at a high voltage level, at which point the thyristor conducts without the application of a Gate signal.

Thyristors can be unintentionally triggered into conduction by a reverse over-voltage, high temperature, or a fast rising dv/dt voltage, such as a spike, which is an essential negative attribute. The two outer P-N junctions are now forward biased, but the center N-P junction is reverse biased, if the Anode terminal is made positive with respect to the Cathode. As a result, forward current is likewise halted. When a positive current is injected into the base of the NPN transistor TR2, the collector current in the base of transistor TR1 flows. As a result, a collector current flows in the PNP transistor TR1, increasing the base current of

TR2, and so on.

Typical Thyristor

Because the two transistors are coupled in an endless regenerative feedback loop, they compel each other to conduct to saturation very quickly. The current flowing through the device between the Anode and the Cathode is limited only by the resistance of the external circuit once it has been triggered into conduction, as the forward resistance of the device when conducting can be very low (less than 1Ω), resulting in low voltage drop across it and power loss.

We observed that in its "OFF" state, a thyristor blocks current in both directions of an AC supply, but that it can be turned "ON" and made to act like a normal rectifying diode by applying a positive current to the base of the transistor, TR2, which is called the "Gate" terminal for a silicon controlled rectifier.

I-V Characteristics Curves of Thyristor

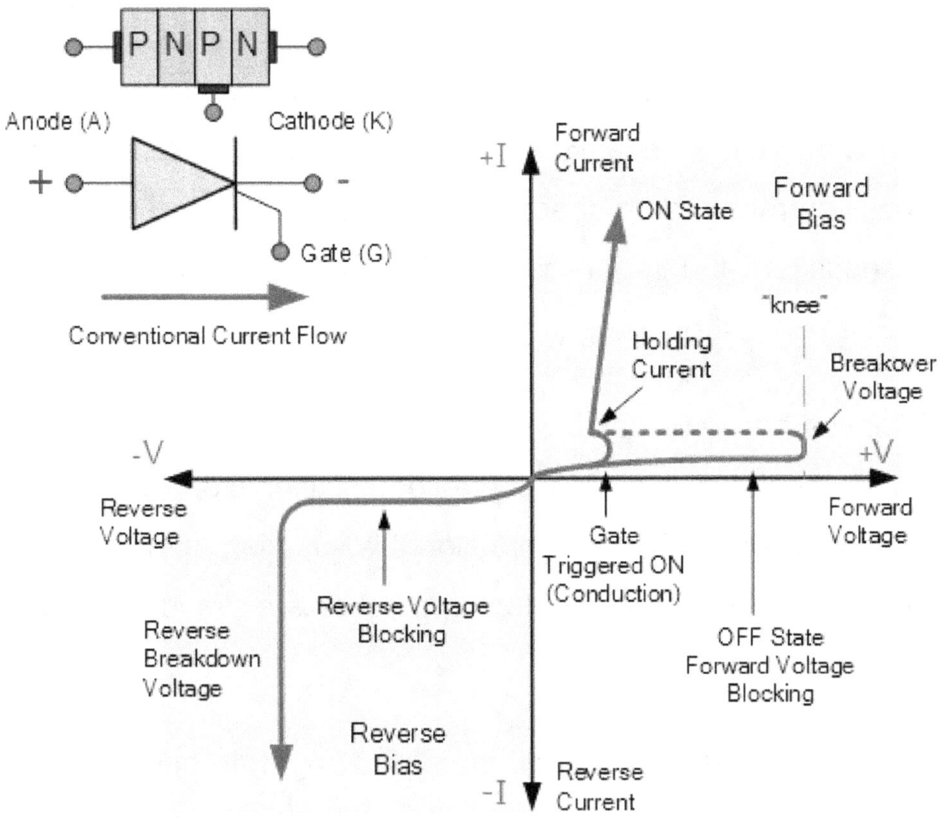

The gate signal loses all control after the thyristor is turned "ON" and passing current in the forward direction (anode positive) due to the regenerative latching action of the two internal transistors. Because the thyristor is already conducting and fully-ON, any gate signals or pulses applied after regeneration has begun would have no effect.

The SCR, unlike the transistor, cannot be biased to remain in an active zone along a load line between blocking and saturation states. Because conduction is controlled internally, the magnitude and length of the gate "turn-on" pulse have no impact on the device's operation. The gadget will conduct if a transient gate pulse is applied to it, and it will remain permanently "ON" even if the gate signal is removed.

Therefore, the thyristor can also be thought of as a *Bistable Latch* having two stable states "OFF" or "ON". This is because with no gate signal applied, a silicon controlled rectifier blocks current in both directions of an AC waveform, and once it is triggered into conduction, the regenerative latching action means that it cannot be turned "OFF" again just by using its Gate.

As a result, the thyristor can be viewed as a Bistable Latch with two stable states: "OFF" and "ON." This is because a silicon controlled rectifier stops current in both directions of an AC waveform when no gate signal is supplied, and once it is triggered into conduction, the regenerative latching mechanism prevents it from being turned "OFF" simply by utilizing its Gate.

Once the thyristor has self-latched into its "ON" state and begun to pass current, it can only be turned "OFF" by either completely removing the supply voltage and thus the Anode (I_A) current, or by reducing the Anode to Cathode current by some external means (for example, by opening a switch) to below a value known as the "minimum holding current (I_H)".

Before a forward voltage is given to the device without it immediately self-conducting, the anode current must be lowered below this minimum holding level long enough for the thyristors internal latching pn-junctions to recover their blocking condition. Clearly, for a thyristor to conduct in the first place, the Anode current, also known as the load current, I_L, must be greater than the holding current, I_H value.

Because the thyristor has the ability to turn "OFF" whenever the Anode current is reduced below this minimum holding value, when used on a sinusoidal AC supply, the SCR will automatically turn "OFF" at some value near the half-cycle cross over point, and will remain "OFF" until the next Gate trigger pulse is applied, as we now know. As this polarity of an AC sinusoidal voltage changes from positive to negative every half-cycle, the thyristor can turn "OFF" at the 180o zero point of the positive waveform. This is referred to as "natural commutation," and it is a key feature of the silicon controlled rectifier.

Thyristors used in circuits fed from DC supplies cannot experience this natural commutation because the DC supply voltage is constant, hence another method of turning the thyristor "OFF" at the right moment must be given because once triggered, it will continue to conduct.

Natural commutation happens every half cycle in AC sinusoidal circuits. The thyristor is thus forward biased (anode positive) during the positive half cycle of an AC sinusoidal waveform and can be triggered "ON" using a Gate signal or pulse. The Anode becomes negative throughout the negative half cycle, whereas the Cathode remains positive. Even if a Gate signal is provided, this voltage reverse biases the thyristor, preventing it from conducting.

The thyristor can be triggered into conduction till the conclusion of the positive half cycle by applying a Gate signal at the right point during the positive half of an AC waveform. Thus, phase control (as it is known) can be used to trigger the thyristor at any position along the positive half of the AC waveform, and power control of AC systems is one of the numerous applications of a Silicon Controlled Rectifier, as shown.

Phase Control of Thyristor

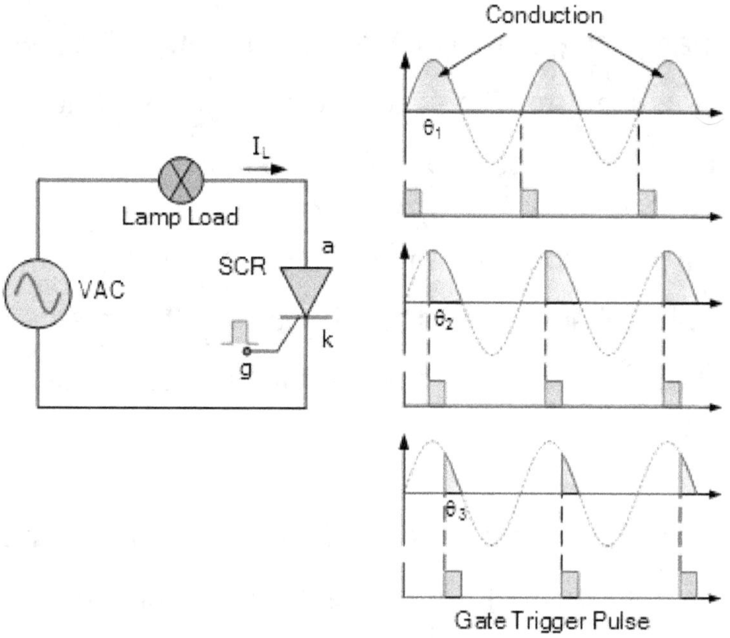

The SCR is "OFF" at the start of each positive half-cycle.

The SCR is triggered into conduction when the gate pulse is applied, and it remains fully latched "ON" for the duration of the positive cycle.The load (a light) will be "ON" for the complete positive cycle of the AC waveform (half-wave rectified AC) at a high average voltage of 0.318 x Vp if the thyristor is activated at the beginning of the half-cycle (= 0°).The lamp is lighted for shorter duration when the gate trigger pulse is applied along the half cycle (= 0° to 90°), and the average voltage given to the lamp is proportionally smaller, decreasing its brightness.

In this way, a silicon controlled rectifier can be used as an AC light dimmer and in a range of other AC power applications like AC motor-speed control, temperature control systems, and power regulator circuits, among others.

We've seen that, regardless of the Gate signal, a thyristor is simply a half-wave device that conducts in only the positive half of the cycle when the Anode is positive and stops current flow like a diode when the Anode is negative. However, there are other semiconductor devices that fall under the "Thyristor" umbrella that can conduct in both directions, are full-wave devices, or can be turned "OFF" by the Gate signal.

"Gate Turn-OFF Thyristors" (GTO), "Static Induction Thyristors" (SITH), "MOS Controlled Thyristors" (MCT), "Silicon Controlled Switch" (SCS), "Triode Thyristors" (TRIAC), and "Light Activated Thyristors" (LASCR) are just a few examples, with all of these devices available in a variety of voltage and current

Thyristor in A Brief

- Thyristors, or Silicon Controlled Rectifiers, are three-junction PNPN semiconductor devices that can be thought of as two interconnected transistors that can be used to switch high electrical loads.
- A single pulse of positive current delivered to their Gate terminal can latch them "ON," and they will stay "ON" eternally until the Anode to Cathode current falls below their minimum latching level.

Thyristor's Static Features

- Thyristors are semiconductor devices that can only transition between states.
- A small Gate current regulates a larger Anode current in a thyristor, which is a current-operated device.

➤ Only conducts current when the Gate is forward biased and a triggering current is applied.

➤ When the thyristor is turned "ON," it operates like a rectifying diode.

➤ To maintain conduction, the anode current must be larger than the holding current.

➤ When reverse biased, regardless of whether gate current is applied, current flow is blocked.

➤ If the anode current is greater than the latching current, the device will be latched "ON" and continue to conduct even if the gate current is no longer delivered.

➤ Thyristors are high-speed switches that can be used to replace electromechanical relays in many circuits since they don't have any moving components, don't cause contact arcing, and don't get dirty or corroded.

➤ Thyristors can be used to adjust the mean value of an AC load current without dissipating enormous quantities of power, in addition to simply switching large currents "ON" and "OFF."

➤ Electric lighting, heaters, and motor speed control are all examples of thyristor power control.

CHAPTER-2: CIRCUIT OF THYRISTOR

Thyristors are solid-state devices with a rapid speed of operation that can be used to control motors, heaters, and lamps. For regenerative latching to occur, we inject a brief trigger pulse of current (not a continuous current) into the Gate, (G) terminal while the thyristor is in its forward orientation, that is, the Anode, (A) is positive with respect to the Cathode, (K).

This trigger pulse can be as short as a few microseconds, but the longer the Gate pulse is delivered, the faster the internal avalanche breaks down and the faster the thyristor turns "ON," but the maximum Gate current must not be exceeded. The voltage drop across the thyristor, Anode to Cathode, is fairly consistent at roughly 1.0V once triggered and fully conducting for all values of Anode current up to its rated value. But keep in mind that once a Thyristor begins to conduct, it will continue to conduct even if the Gate signal is absent, until the Anode current falls below the device's holding current (I_H), at which point it will automatically turn "OFF." Thyristors cannot be utilized for amplification or controlled switching, unlike bipolar transistors and FETs.

Semiconductor devices like thyristors don't have amplification capability. They are specifically intended for use in high-power switching applications. Thyristors can only function in switching mode, operating as an open or closed switch. A thyristor will always conduct (pass current) once it is triggered into conduction by its gate terminal. As a result, in DC circuits and some highly inductive AC circuits, a separate switch or turn off circuit is required to artificially limit the current.

Circuit with DC Thyristor

The thyristor can be used as a DC switch to regulate bigger DC currents and loads when coupled to a direct current DC supply. When used as a switch, the Thyristor acts like an electrical latch since it remains in the "ON" state unless manually reset. Have a look at the DC thyristor circuit below.

Switching Circuit with DC Thyristors

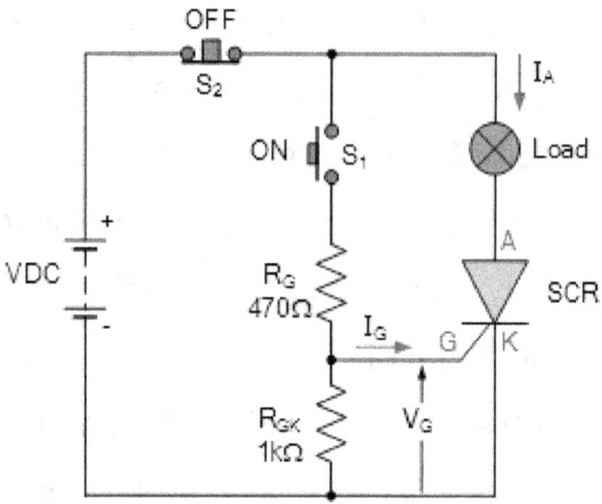

The thyristor is employed as a switch in this simple "on-off" thyristor firing circuit to control a lamp, but it could also be used as an on-off control circuit for a motor, heater, or other DC load. The thyristor is forward biased and is pushed into conduction by quickly shutting the normally open "ON" push button, S1, which allows current to flow into the Gate by connecting the Gate terminal to the DC supply via the Gate resistor, RG. The thyristor may not trigger if the value of RG is set too high in relation to the supply voltage.

After the circuit is turned "ON," it latches and remains "ON" even when the push button is withdrawn, as long as the load current exceeds the thyristor's latching current. Additional actions of the push button, S1, will have no effect on the state of the circuit because the Gate has lost all control once it has been "latched." The thyristor is now completely "ON" (conducting), allowing full load circuit current to flow forward through the device and back to the battery supply.

One of the primary benefits of employing a thyristor as a switch in a DC circuit is its extremely high current gain. Because a modest Gate current may control a much greater Anode current, the thyristor is a current-operated device. The Gate-cathode resistor RGK is typically included to lessen the Gate's sensitivity and enhance its dv/dt capacity, hence preventing the device from being falsely triggered. Because the thyristor has self-latched into the "ON" state, the circuit can only be reset by cutting the power supply and lowering the anode current to less than the thyristor's minimum holding current (I_H).

S2 breaks the circuit by pressing the normally closed "OFF" push button, lowering the circuit current flowing through the Thyristor to zero and causing it to turn "OFF" until another Gate signal is applied. The mechanical normally-closed "OFF" switch S2 must be large enough to handle the circuit power flowing through both the thyristor and the light when the contacts are opened, which is one of the downsides of this DC thyristor circuit design. If this is the case, we could simply substitute a huge mechanical switch for the thyristor. To circumvent this issue and avoid the requirement for a larger, more durable "OFF" switch, connect the switch in parallel with the thyristor as indicated.

Alternative DC Circuit with a Thyristor

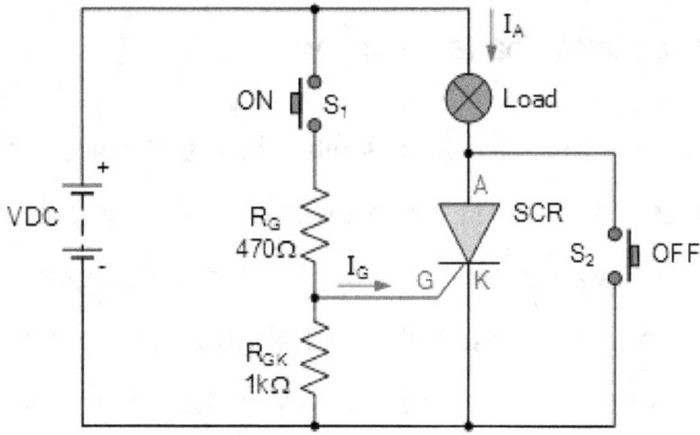

In this case, the necessary terminal voltage and Gate pulse signal are applied to the thyristor switch as before, but the bigger normally-closed switch from the previous circuit has been replaced by a smaller normally-open switch in parallel with the thyristor. When switch S2 is activated, it creates a temporary short circuit between the Anode and Cathode of the thyristors, preventing the device from conducting by lowering the holding current to below its minimum value.

AC Circuit with a Thyristor

The thyristor behaves differently than the prior DC-connected circuit when linked to an alternating current AC supply. Because alternating current switches polarity on a regular basis, any thyristor used in an alternating current circuit will be reverse-biased, forcing it to turn "OFF" during one-half of each cycle. Consider the following AC thyristor circuit.

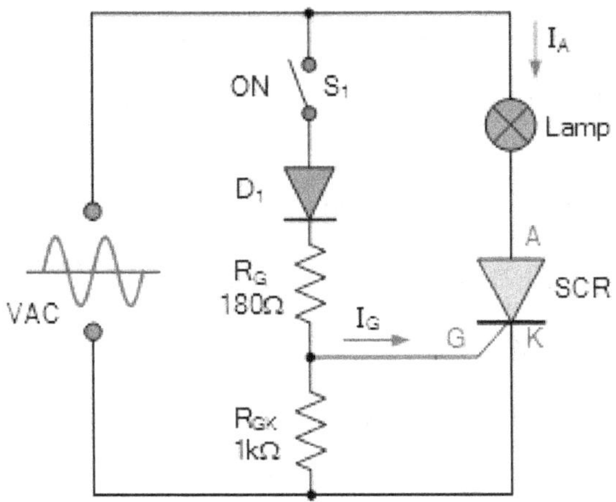

Except for the lack of an additional "OFF" switch and the inclusion of diode D1, which prevents reverse bias from being applied to the Gate, the above thyristor firing circuit is similar to the DC SCR circuit. The device is forward biased during the positive half-cycle of the sinusoidal waveform, but with switch S1 open, zero gate current is delivered to the thyristor, and it remains "OFF." The device is reverse biased on the negative half-cycle and will remain "OFF" regardless of the state of switch S1.

If switch S1 is closed, the thyristor is entirely "OFF" at the start of each positive half-cycle, but there will be enough positive trigger voltage and hence current present at the Gate shortly after to turn the thyristor and the lamp "ON." The thyristor is now latched-"ON" for the duration of the positive half-cycle, and will turn "OFF" when the positive half-cycle ends and the Anode current falls below the holding current value. The device is entirely "OFF" for the next negative half-cycle until the next positive half-cycle, when the process repeats itself and the thyristor conducts again as long as the switch is closed.

Since the thyristor functions like a rectifying diode and conducts current only during the positive half-cycles when it is forward biased, the lamp will only receive half of the available power from the AC source under this scenario. Until the switch is opened, the thyristor continues to deliver 50% power to the lamp.

The device would only conduct for one half of a positive half-cycle if switch S1 could be turned on and off quickly enough so that the thyristor received its Gate signal at the "peak" (90°) point of each positive half-cycle. To put it another way, conduction would only occur during one-half of one-half of a sine wave, resulting in the lamp receiving "one-fourth" or a quarter of the total power available from the AC source.

The Thyristor could be configured to supply any proportion of power to the load between 0 and 50% by precisely changing the timing relationship between the Gate pulse and the positive half-cycle. Obviously, this circuit layout cannot give more than 50% power to the light because it is reverse biased and cannot conduct during the negative half-cycles. Consider the following circuit.

Half Wave Phase Control

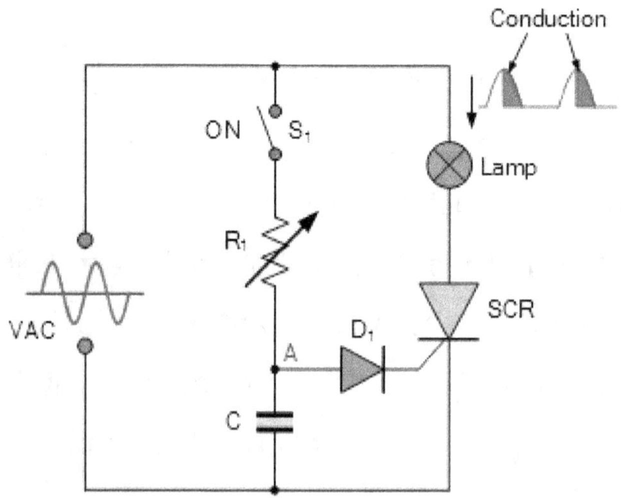

The most widely used form of thyristor AC power control is phase control, and a simple AC phase-control circuit can be built as shown above. The gate voltage of the thyristor is acquired from the RC charging circuit via the trigger diode, D1.

When the thyristor is forward biased during the positive half-cycle, capacitor C charges up via resistor R1 in response to the AC supply voltage. When the voltage at point A rises sufficiently to make the trigger diode D1 to conduct, the capacitor discharges into the Gate of the thyristor, turning it "ON." The variable resistor, R1, controls the time duration in the positive half of the cycle during which conduction begins.

Raising the value of R1 delays the triggering voltage and current provided to the thyristors Gate, causing a lag in the device's conduction time. As a result, the fraction of the half-cycle over which the device conducts can be altered between 0 and 180o, allowing the lamp's average power to be adjusted. Since the thyristor is a unidirectional device, it can only supply a maximum of 50% power during each positive half-cycle.

Employing "thyristors," you can get 100 percent full-wave AC control with various methods. One option is to employ a single thyristor in a diode bridge rectifier circuit that converts AC to a unidirectional current through the thyristor, while the more frequent method is to use two thyristors in inverse parallel. A single Triac is a more practical technique because it may be triggered in both directions, making it suited for AC switching applications.

CHAPTER-3: TRIAC

A triac is a solid-state device that can switch and control AC electricity in both directions of a sinusoidal waveform at fast speeds. Thyristors are solid-state devices that can be used to control lamps, motors, and heaters, among other things.

However, like a diode, a thyristor is a unidirectional device, meaning it only transfers current in one way, from Anode to Cathode. This is one of the drawbacks of utilizing a thyristor to operate such circuits. This "one-way" switching characteristic may be acceptable for DC switching circuits since once triggered, all DC power is sent directly to the load. This unidirectional switching, however, may be an issue with sinusoidal AC switching circuits because it only conducts for one half of the cycle (like a half-wave rectifier) when the Anode is positive, regardless of what the Gate signal is doing. Then, for AC operation, a thyristor delivers half of the power to the load.

To achieve full-wave power control, we could use a single thyristor inside a full-wave bridge rectifier that triggers on each positive half-wave, or we could use two thyristors in inverse parallel (back-to-back) as shown below, but this increases the switching circuit's complexity and number of components.

Configurations of Thyristor

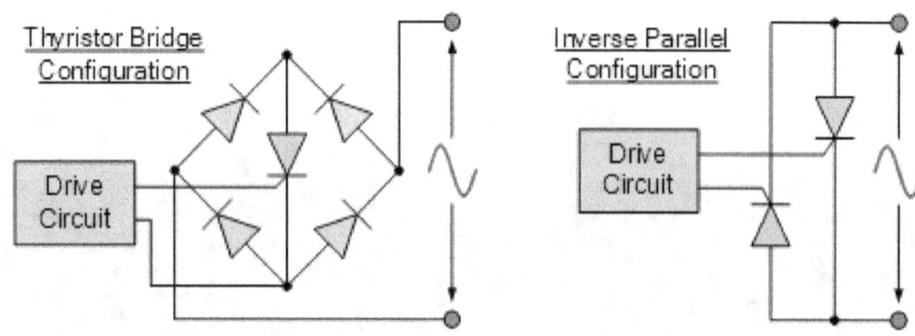

A "Triode AC Switch," or Triac for short, is a kind of semiconductor device that is also a member of the thyristor family and can be employed as a solid state power switching device.The fact that a triac is a "bidirectional" switching device gives it an advantage over a silicon controlled rectifier (SCR). A Triac can be triggered into conduction by both positive and negative voltages given to its Anode, as well as positive and negative trigger pulses applied to its Gate terminal, making it a two-quadrant switching Gate controlled device.

A Triac operates similarly to two conventional thyristors connected in inverse parallel (back-to-back) with regard to one other, with the two thyristors sharing a common Gate terminal all within a single three-terminal package. Since a triac conducts in both directions of a sinusoidal waveform, the concept of an anode terminal and a cathode terminal used to identify a thyristor's main power terminals is replaced with identifications of: MT1, for Main Terminal 1, and MT2, for Main Terminal 2, with the Gate Terminal G referenced the same.

Similar to the gate-cathode relationship of the thyristor or the base-emitter relationship of the transistor, the triac gate terminal is related with the MT1 terminal in most AC switching applications. Below is a diagram of a Triac's construction, P-N doping, and schematic symbol.

Symbol and Construction of Triac

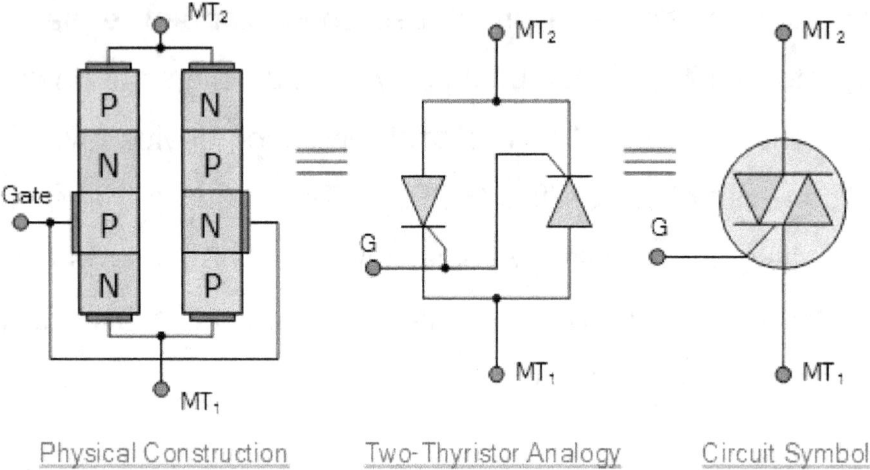

Physical Construction   Two-Thyristor Analogy   Circuit Symbol

A four-layer, three-terminal bidirectional device with PNPN in the positive direction and NPNP in the negative direction that acts as an open-circuit switch in its "OFF" state is called a "triac". But unlike a conventional thyristor, the triac can conduct current in either direction when triggered by a single gate pulse. The following are the four possible triggering modes of operation for a triac.

- I + Mode = MT2 current positive (+ve), Gate current positive (+ve)
- I – Mode = MT2 current positive (+ve), Gate current negative (-ve)
- III + Mode = MT2 current negative (-ve), Gate current positive (+ve)
- III – Mode = MT2 current negative (-ve), Gate current negative (-ve)

The triac's I-V characteristics curves depict these four modes of operation.

I-V Characteristics Curves of a Triac

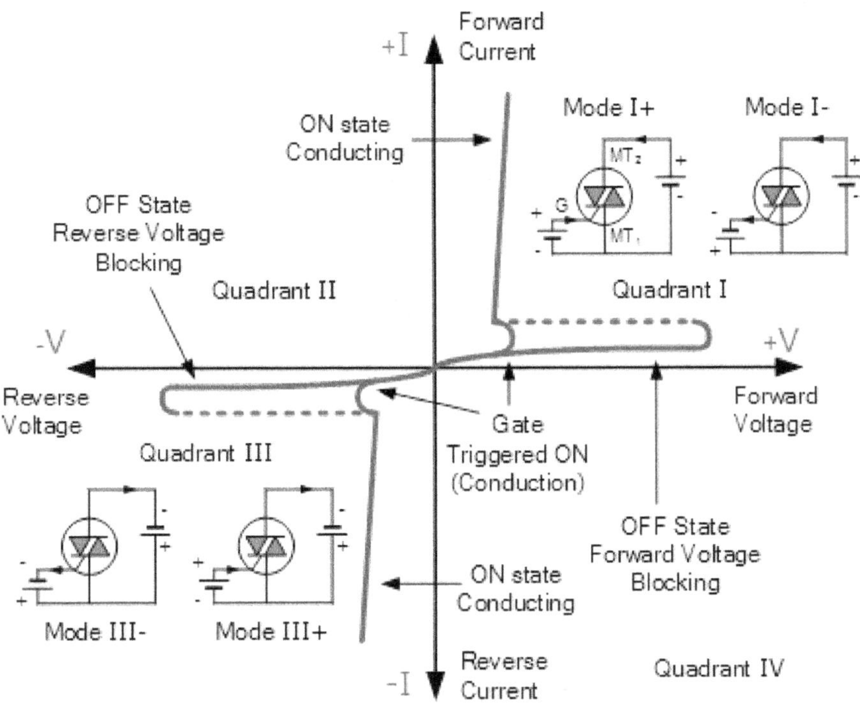

A positive gate current, labeled above as mode I+, normally triggers the triac into conduction in Quadrant I. It can, however, be triggered by a negative gate current, mode I–. Likewise, in Quadrant <III, triggering with a negative gate current, $-I_G$ is also common, mode III– and mode III+. Modes I– and III+, on the other hand, are less sensitive designs that require a higher gate current to activate than the more usual triac triggering modes of I+ and III–.

In addition, triacs, like silicon controlled rectifiers (SCRs), require a minimum holding current IH to maintain conduction at the waveform's cross over point. Even though the two thyristors are merged into a single triac device, they nevertheless display unique electrical characteristics such as differing breakdown voltages, holding currents, and trigger voltage levels, much like a single SCR device.

Uses of Triac

Since the triac may be switched "ON" by either a positive or negative Gate pulse, independent of the polarity of the AC supply at the time, it is the most often used semiconductor device for switching and power regulation of AC systems. With the extremely basic triac switching circuit shown below, the triac is perfect for controlling a lamp or AC motor load.

Basic Switching Circuit of a Triac

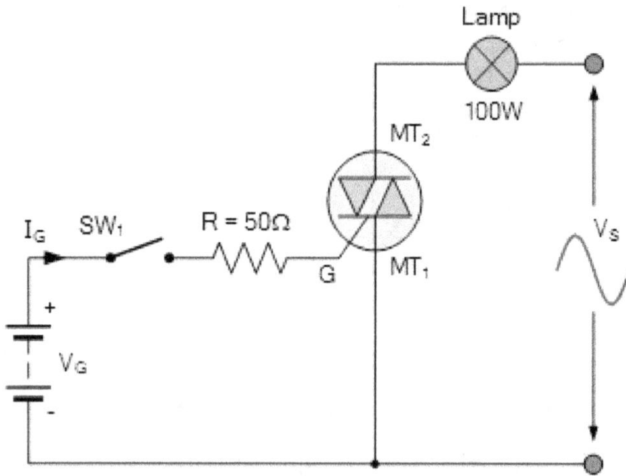

A basic DC triggered triac power switching circuit is shown in the diagram above. The bulb is "OFF" when switch SW1 is open since no current passes into the triac's Gate. When SW1 is closed, gate current from the battery supply VG is applied to the triac via resistor R, and the triac is forced into full conduction, operating as a closed switch, drawing full power from the sinusoidal supply. The triac is continuously gated in modes I+ and III+ regardless of the polarity of terminal MT2, because the battery sends a positive Gate current to the triac anytime switch SW1 is closed.

Obviously, the difficulty with this simple triac switching circuit is that it requires a separate positive or negative Gate supply to initiate conduction. However, we can alternatively use the AC supply voltage as the gate triggering voltage to trigger the triac. Consider the following circuit:

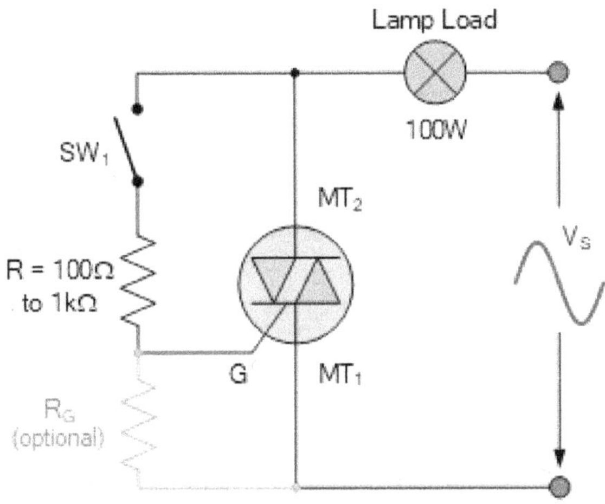

The circuit illustrates a triac being used as a simple static AC power switch with a "ON"-"OFF" function, similar to the DC circuit. The triac works as an open switch when switch SW1 is open, and the lamp draws zero current.

When SW1 is closed, the triac is gated "ON" via current limiting resistor R and self-latches shortly after the start of each half-cycle, allowing full power to be sent to the light load. The triac automatically unlatches at the end of each AC half-cycle as the instantaneous supply voltage falls to zero, causing the load current to briefly fall to zero, but re-latches using the opposite thyristor half on the next half-cycle as long as the switch remains closed because the supply is sinusoidal AC. Since both sides of the sine wave are controlled, this sort of switching control is referred known as full-wave control. As the triac is effectively two back-to-back coupled SCRs, we can extend the functionality of this triac switching circuit by changing the way the gate is triggered, as shown below.

Modification of Switching Circuit with Triac

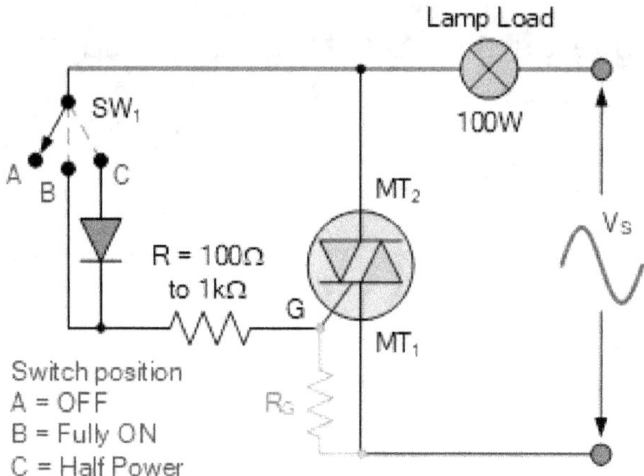

There is no gate current and the bulb is "OFF" if switch SW1 is open at position A, as shown above. When the switch is switched to position B, the gate current flows at the same rate as before, and the lamp draws full power because the triac functions in modes I+ and III−.

When the switch is connected to position C, however, the diode will prevent the gate from triggering when MT2 is negative since the diode is reverse biased. As a result, the triac will only conduct on positive half-cycles while operating in mode I+, and the lamp will only light half-power. The load is then Off, Half Power, or Fully ON, depending on the position of the switch.

Triac switching circuit with phase control

For both the positive and negative sides of the input waveform, a typical type of triac switching circuit uses phase control to adjust the amount of voltage and thus power applied to a load, in this case a motor. Since the voltage can be varied from zero to the entire applied voltage as illustrated, this sort of AC motor speed control provides fully variable and linear control.

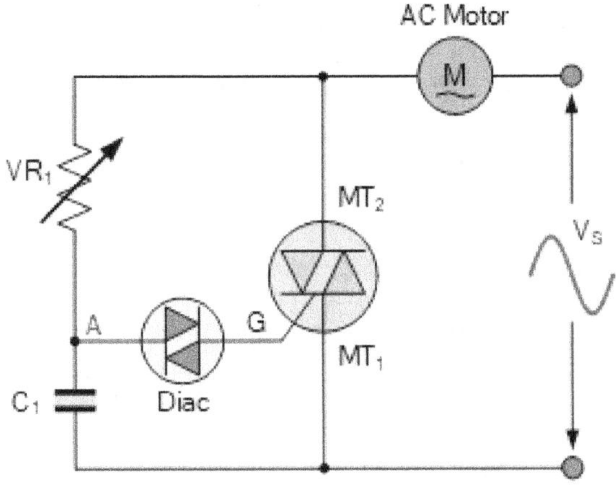

The triac is connected in series with the motor across an AC sinusoidal supply in this basic phase triggering circuit. VR1 is a variable resistor that controls the degree of phase shift on the gate of the triac, which controls the amount of voltage provided to the motor by turning it on at different points during the AC cycle.

The triggering voltage of the triac is derived from the VR1 – C1 combination through the Diac (The diac is a bidirectional semiconductor device that helps provide a sharp trigger current pulse to fully turn-ON the triac).

C1 charges up via the variable resistor, VR1, at the start of each cycle. This process is repeated until the voltage across C1 is sufficient to cause the diac to conduct, allowing the capacitor, C1, to discharge into the gate of the triac, turning it "ON."

When the triac is triggered into conduction and saturates, it essentially shorts out the gate triggering phase control circuit that is linked in parallel across it, and the triac assumes control for the rest of the half-cycle. As previously stated, the triac automatically turns off at the conclusion of the half-cycle, and the VR1 – C1 triggering mechanism begins again on the next half-cycle.

Since each switching mode of operation, such as I+ and III–, necessitates different amounts of gate current, a triac is asymmetrical, meaning it may not trigger at the same position for each positive and negative half cycle.

This simple triac speed control circuit is useful for controlling not just AC motors, but also lighting dimmers and electrical heaters, and is quite similar to a common triac light dimmer. A commercial triac dimmer, on the other hand, should not be utilized as a motor speed controller since triac light dimmers are designed for resistive loads only, such as incandescent lamps.

Triac at a glance:

> A "Triac," like the SCR, is a 4-layer, 3-terminal thyristor device.
> In either way, the Triac can be triggered into conduction.
> A Triac has four different triggering modes, two of which are recommended.
> When used appropriately to regulate resistive type loads such as incandescent lamps, heaters, or small universal motors frequently found in portable power tools and small appliances, electrical AC power management utilizing a Triac is incredibly successful.
> However, keep in mind that these devices can be connected to the mains AC power supply, therefore circuit testing should be performed while the power control device is detached from the mains power supply. Please keep in mind that safety comes first!

CHAPTER-4: INSULATED GATE BIPOLAR TRANSISTOR (IGBT)

For its use in power supply and motor control circuits, the IGBT is a power switching transistor that combines the advantages of MOSFETs and BJTs. The Insulated Gate Bipolar Transistor, or IGBT for short, is a hybrid of a Bipolar Junction Transistor (BJT) and a Field Effect Transistor (MOSFET), making it perfect for use as a semiconductor switching device.

The IGBT Transistor combines the best features of these two common transistor types, combining the high input impedance and high switching speeds of a MOSFET with the low saturation voltage of a bipolar transistor to create a new type of transistor switching device capable of handling large collector-emitter currents with virtually no gate current drive.

Typical IGBT

The Insulated Gate Bipolar Transistor (IGBT) combines MOSFET insulated gate technology with the output performance characteristics of a normal bipolar transistor (hence the second part of its name). The "IGBT Transistor" has the output switching and conduction characteristics of a bipolar transistor but is voltage-controlled like a MOSFET as a result of this hybrid arrangement.

IGBTs are primarily employed in power electronics applications such as inverters, converters, and power supplies where the solid-state switching device demands are not fully covered by power bipolars and power MOSFETs. High-voltage and high-current bipolars are available, but their switching rates are slow, whereas power MOSFETs may have faster switching speeds, but high-voltage and high-current devices are expensive and difficult to achieve.

The insulated gate bipolar transistor device has an advantage over a BJT or MOSFET in that it provides more power gain than a regular bipolar type transistor while operating at a higher voltage and with reduced input losses. As illustrated below, it is a FET integrated with a bipolar transistor in a Darlington type design.

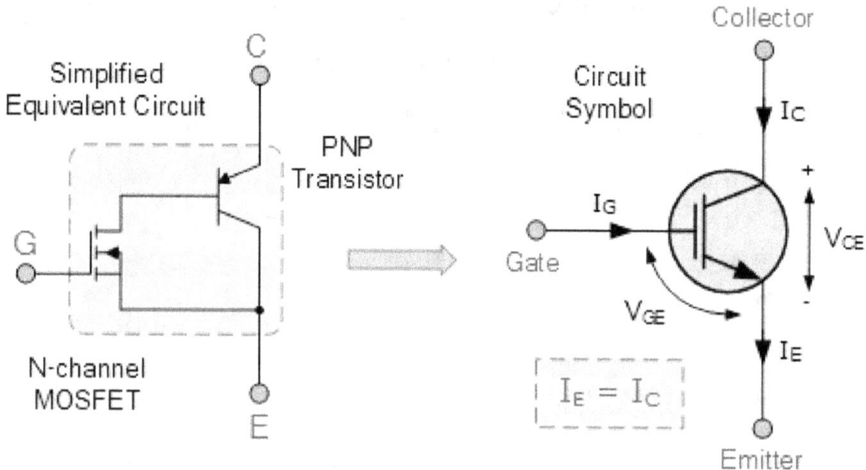

The insulated gate bipolar transistor is a three-terminal trans conductance device that combines an insulated gate N-channel MOSFET input and a PNP bipolar transistor output coupled in a Darlington configuration. As a result, the terminals are named Collector, Emitter, and Gate. Two of its terminals (C-E) are connected to the conductance path through which current flows, while its third terminal (G) controls the device. The insulated gate bipolar transistor's amplification is proportional to the ratio of its output signal to its input signal.The amount of gain in a standard bipolar junction transistor (BJT) is approximately equal to the ratio of the output current to the input current, known as Beta.

There is no input current in a metal oxide semiconductor field effect transistor, or MOSFET, because the gate is isolated from the main current carrying channel. As a result, the gain of a FET is equal to the ratio of output current change to input voltage change, making it a trans conductance device, as is true of the IGBT.

The IGBT can thus be treated as a power BJT with a MOSFET providing the base current. The Insulated Gate Bipolar Transistor, like the BJT or MOSFET type transistors, can be utilized in tiny signal amplifier circuits. Since the IGBT combines the low conduction loss of a BJT with the fast switching speed of a power MOSFET, it is an ideal solid state switch for use in power electronics.

In addition, compared to a comparable MOSFET, the IGBT has a substantially lower "on-state" resistance, or RON. As a result, for a given switching current, the I2R drop across the bipolar output structure is substantially lower. The IGBT transistor functions similarly to a power MOSFET in terms of forward blocking. The insulated gate bipolar transistor has similar voltage and current ratings to the bipolar transistor when employed as a static controlled switch.

The existence of an isolated gate in an IGBT, on the other hand, makes it considerably easier to drive than a BJT since much less drive power is required. The Gate terminal of an insulated gate bipolar transistor is easily activated and deactivated to make it "ON" or "OFF."

A positive input voltage signal across the Gate and the Emitter will keep the device "ON," while a zero or slightly negative input gate signal will cause it to turn "OFF," much like a bipolar transistor or eMOSFET. Another advantage of the

IGBT is that its on-state channel resistance is substantially lower than that of a typical MOSFET.

Characteristics of IGBT

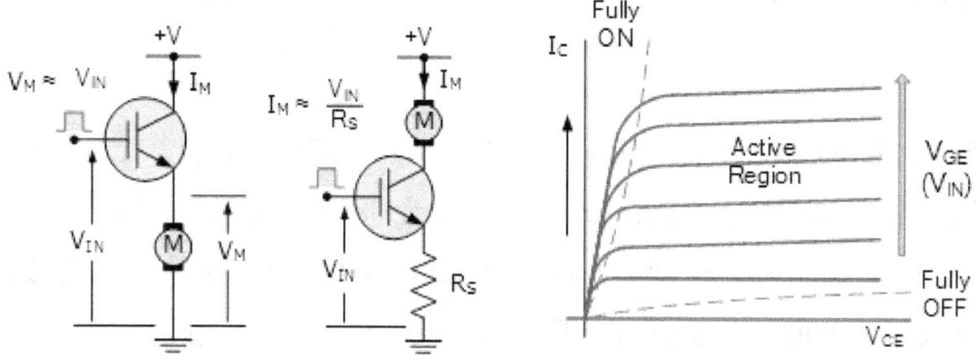

Since the IGBT is a voltage-controlled device, it needs a minimal voltage on the Gate to keep conduction through the device, as opposed to BJTs, which require that the Base current be supplied continually in sufficient quantity to keep saturation.

Furthermore, the IGBT is a unidirectional device, which indicates it can only switch current in the "forward direction," that is, from Collector to Emitter, as opposed to MOSFETs, which may switch current in both directions (controlled in the forward direction and uncontrolled in the reverse direction).

The operating principle and gate drive circuits of an insulated gate bipolar transistor are quite similar to those of an N-channel power MOSFET. The key difference is that when current flows through the device in its "ON" state, the resistance offered by the main conducting channel is substantially lower in the IGBT. As a result, the current ratings are significantly higher as compared to an equivalent power MOSFET.

High voltage capability, low ON-resistance, ease of drive, relatively fast switching speeds, and zero gate drive current make the Insulated Gate Bipolar Transistor a good choice for moderate speed, high voltage applications such as pulse-width modulated (PWM), variable speed control, switch-mode power supplies, or solar powered DC-AC inverter and frequency converter applications. The following table shows a general comparison of BJTs, MOSFETs, and IGBTs.

Table of IGBT Comparison

Characteristics of Device	Power Bipolar	Power MOSFET	IGBT
Rated Voltage	High <1kV	High <1kV	Very High >1kV
Rated Current	High <500A	Low <200A	High >500A
Input Drive	Current, h_{FE} 20-200	Voltage, V_{GS} 3-10V	Voltage, V_{GE} 4-8V
Input Impedance	Low	High	High
Output Impedance	Low	Medium	Low
Switching Speed	Slow (uS)	Fast (nS)	Medium
Cost	Low	Medium	High

The Insulated Gate Bipolar Transistor (IGBT) is a semiconductor switching device with the output properties of a bipolar junction transistor (BJT) but the control characteristics of a metal oxide field effect transistor (MOSFET).

One of the key advantages of the IGBT transistor is how easily it can be turned "on" by applying a positive gate voltage or turned "off" by making the gate signal zero or slightly negative, allowing it to be utilized in a wide range of switching applications. For usage in power amplifiers, it can also be driven in its linear active area.

The Insulated Gate Bipolar Transistor is ideal for driving inductive loads such as coil windings, electromagnets, and DC motors because of its decreased on-state resistance and conduction losses, as well as its ability to switch high voltages at high frequencies without damage.

CHAPTER-5: DIAC

The Diac is a bidirectional semiconductor device with two junctions which is designed in such a way that the AC voltage across it surpasses a particular level, allowing current to flow in either direction. The diode AC switch, or Diac for short, is a solid-state, three-layer, two-junction semiconductor device, however unlike the transistor, it lacks a base connection, making it a two-terminal device with A1 and A2. Diacs are electronic components that have no control or amplification but can conduct current from either polarity of a suitable AC voltage supply, similar to a bidirectional switching diode.

In ON-OFF switching applications, we observed that these devices are controlled by simple circuits producing steady state gate currents.

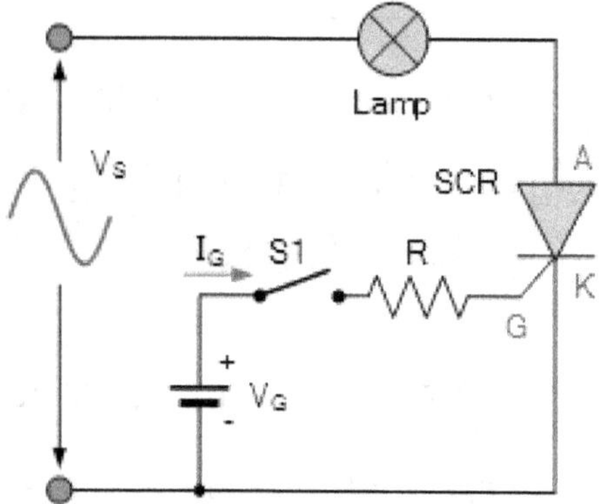

The bulb is "OFF" while switch S1 is open since no gate current passes. When switch S1 is closed, gate current IG flows, and the SCR conducts exclusively on the positive half cycles, indicating that it is in quadrant I. When the SCR is gated

"ON," it will only switch "OFF" when the supply voltage falls to a level where the anode current, I_A, is less than the holding current, I_H.

A short pulse of gate current at a pre-set trigger point can be provided to allow conduction of the SCR to occur throughout portion of the half-cycle only if we wanted to adjust the mean value of the lamp current rather than just switch it "ON" or "OFF". Therefore, by varying the delay time, T, between the start of the cycle and the trigger point, the mean value of the lamp current could be adjusted. The term "phase control" is used to describe this procedure. However, two things are required for phase control. The first is a variable phase shift circuit (typically an RC passive circuit), and the second is a trigger circuit or device that may generate the needed gate pulse when the delayed waveform reaches a particular level.

The Diac is one solid-state semiconductor device that can generate these gate pulses. The diac is built similarly to a transistor, however it does not have a base connection, allowing it to be used in either polarity in a circuit.

Diacs are typically employed as trigger devices in phase-triggering and variable power control applications because they help generate a sharper and more immediate trigger pulse (rather than a continuously rising ramp voltage) that is used to turn "ON" the primary switching device. The symbol and voltage-current characteristics curves of a diac are shown below.

Symbol and I-V Characteristics of Diac

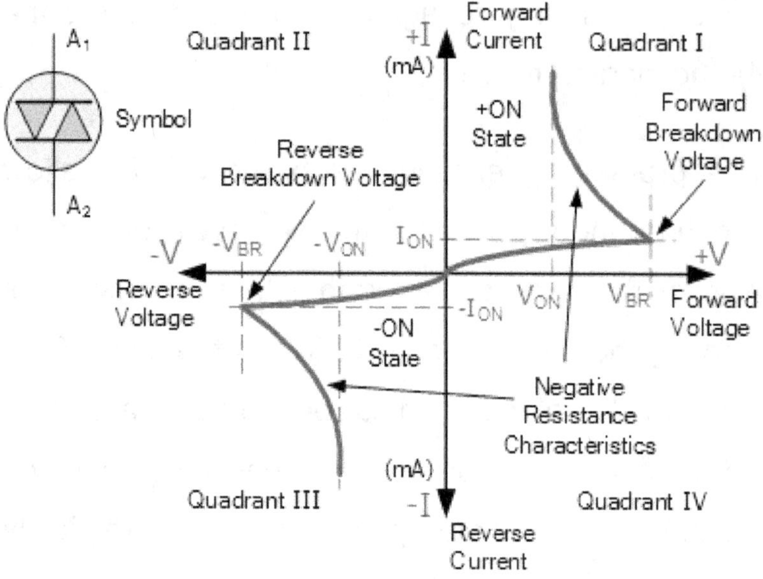

The above diac I-V characteristics curves show that the diac blocks current flow in both directions until the applied voltage is greater than VBR, at which point the device fails and the diac conducts heavily, similar to a zener diode passing a sudden voltage pulse.

The Diacs breakdown voltage or breakover voltage is the VBR point. The voltage across a standard zener diode would remain constant as the current increased. The transistor action in the diac, on the other hand, causes the voltage to decrease as the current increases. Once in the conducting state, the diac's resistance drops to a very low level, allowing a large amount of current to flow. The breakdown voltage of most commonly available diacs, such as the ST2 or DB3, is typically between 25 and 35 volts.

Higher break over voltage ratings, such as 40 volts for the DB4 diac, are available. As shown above, this action gives the diac the characteristic of negative resistance. As the diac is symmetrical, it has the same characteristic for both positive and negative voltages, and it is this negative resistance action that makes it suitable for triggering SCRs or triacs.

Applications of Diac

The diac is commonly used as a solid state triggering device for other semiconductor switching devices, primarily SCRs and triacs, as stated above. The diac is used in conjunction with the triac to provide full-wave control of the AC supply, as shown. Triacs are widely used in applications such as lamp dimmers and motor speed controllers.

Diac AC Phase Control

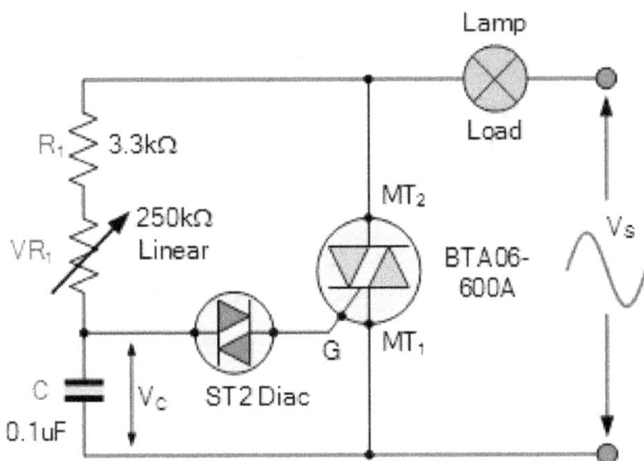

The capacitor, C, is charged through the series combination of the fixed resistor, R1, and the potentiometer, VR1, as the AC supply voltage rises at the start of the cycle, and the voltage across its plates rises.

The diac breaks down when the charging voltage reaches the diac's breakover voltage (about 30 V for the ST2), and the capacitor discharges through the diac.

The discharge generates a burst of current that causes the triac to conduct. VR1, which controls the charging rate of the capacitor, can change the phase angle at which the triac is triggered. When VR1 is at its lowest, R1 limits the gate current to a safe level.

The load current maintains the triac's "ON" state, while the voltage across the resistor–capacitor combination is limited by the triac's "ON" voltage and is maintained until the end of the current half-cycle of the AC supply.

The supply voltage drops to zero at the end of the half cycle, lowering the current through the triac below its holding current, I_H turning it "OFF," and the diode stops conduction. The supply voltage then enters its next half-cycle, the capacitor voltage rises again (this time in the opposite direction), and the triac firing cycle repeats.

Waveform During Conduction by Triac

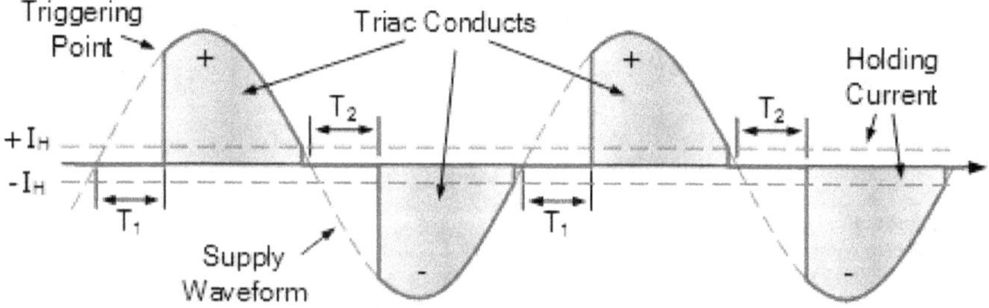

The Diac is a very useful device that can be used to trigger triacs and, due to its negative resistance characteristics, it can switch "ON" quickly once a certain applied voltage level is reached. This means that whenever we use a triac for AC power control, we will also need a separate diac. Fortunately for us, someone somewhere decided to replace the individual diac and triac with a single switching device known as a Quadrac.

Quadracs

Quadracs are essentially Diacs and Triacs fabricated together within a single semiconductor package and are thus referred to as "internally triggered triacs." This bi-directional all-in-one device can be gate controlled using either polarity of the main terminal voltage, allowing it to be used in full-wave phase-control applications such as heater controls, lamp dimmers, and AC motor speed control, among others.

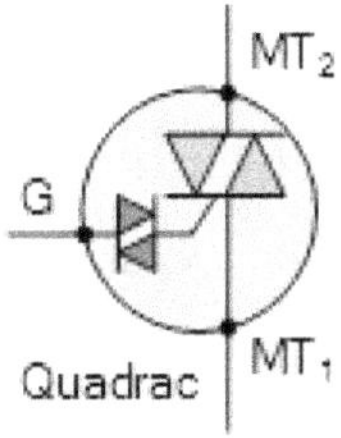

Quadracs, like triacs, are three-terminal semiconductor switching devices denoted by MT2 for main terminal one (usually the anode), MT1 for main terminal two (usually the cathode), and G for the gate terminal. The quadrac is available in a variety of package types depending on the voltage and current switching requirements, the most common being the TO-220 package. The quadrac is intended to be a direct replacement for the majority of triac devices.

Diac in A Brief

Diac is a two-terminal voltage blocking device that can conduct in either direction, like the ST2 or DB3. Diacs have negative resistance characteristics that allow them to quickly switch "ON" when a certain applied voltage level is reached. As the diac is bidirectional, it can be used as a triggering device in phase control and general AC circuits such as light dimmers and motor speed controls when paired with the BTAxx-600A or IRT80 series of switching triacs.

Quadracs are simply triacs with a diac connected internally. Quadracs are bidirectional AC switches that are gate controlled for either polarity of main terminal voltage, similar to triacs.

CHAPTER-6: UNIJUNCTION TRANSISTOR (UJT)

The UJT is a three-terminal semiconductor device with negative resistance and switching characteristics that can be used as a relaxation oscillator in phase control applications. The Unijunction Transistor, or UJT for short, is another solid state three terminal device that can be used in gate pulse, timing circuits, and trigger generator applications to switch and control thyristors and triacs for AC power control type applications.

Unijunction transistors, like diodes, are built from separate P-type and N-type semiconductor materials, resulting in a single (hence the name Uni-Junction) PN-junction within the device's main conducting N-type channel.

Even though the Unijunction Transistor is called a transistor, its switching characteristics are very different from those of a conventional bipolar or field effect transistor because it is used as an ON-OFF switching transistor rather than a signal amplifier. During breakdown, UJTs have unidirectional conductivity and negative impedance characteristics, making them more like a variable voltage divider.

The UJT, like N-channel FETs, is made up of a single solid piece of N-type semiconductor material that forms the main current carrying channel, with Base 2 (B2) and Base 1 (B1) as the outer connections (B1). Along the channel is the third connection, which is confusingly labeled as the Emitter (E).

The unijunction transistor's emitter rectifying p-n junction is created by fusing P-type material into the N-type silicon channel. Although P-channel UJTs with an N-type Emitter terminal are available, they are rarely used. The Emitter junction is placed along the channel, closer to terminal B2 than terminal B1. The UJT symbol contains an arrow pointing towards the base, indicating that the Emitter terminal is positive and the silicon bar is negative. The UJT's symbol, construction, and equivalent circuit are shown below.

Symbol and Construction of Unijunction Transistor

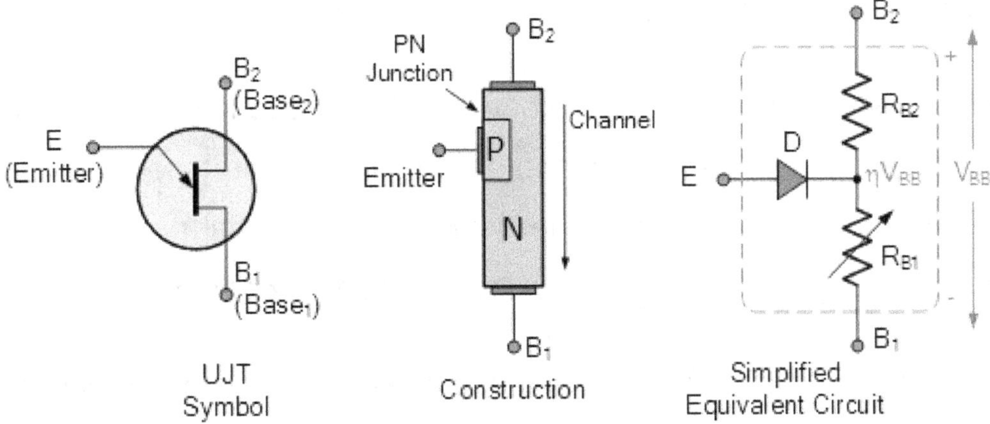

Unijunction transistor's symbol is very close to junction field effect transistor or JFET, with the exception that the Emitter(E) input is represented by a bent arrow. While their ohmic channels are similar, JFETs and UJTs operate very differently and should not be confused.

The N-type channel is basically two resistors RB2 and RB1 in series with an equivalent (ideal) diode, D representing the p-n junction connected to their center point, as shown in the equivalent circuit above. The position of the Emitter p-n junction along the ohmic channel is fixed during manufacturing and cannot be changed.

Resistance RB1 exists between the Emitter, E, and terminal B1, whereas resistance RB2 exists between the Emitter, E, and terminal B2. Since the p-n junction is relatively close to terminal B2 than terminal B1, the resistive value of RB2 is less than RB1.

Silicon bar's total resistance (its Ohmic resistance) is determined by the semiconductor's actual doping level as well as the physical dimensions of the N-type silicon channel, but it can be represented by RBB. When measured with an ohmmeter, the static resistance of most common UJTs, such as the 2N1671, 2N2646, or 2N2647, would typically be in the 4k to 10k range.

These two series resistances form a voltage divider network between the unijunction transistor's two base terminals, and as this channel stretches from B2 to B1, when a voltage is applied across the device, the potential at any point along the channel will be proportional to its position between B2 and B1.

As a result, the magnitude of the voltage gradient is determined by the amount of supply voltage. Terminal B1 is connected to ground in a circuit, and the Emitter is used as the device's input. Assume that a voltage VBB is applied across the UJT between B2 and B1, biasing B2 positive in comparison to B1.

Voltage of Unijunction Transistor Across R_{B1}

$$V_{RB1} = \frac{R_{B1}}{R_{B1} + R_{B2}} \times V_{BB}$$

The resistive ratio of RB1 to RBB shown above for a unijunction transistor is known as the intrinsic stand-off ratio and is denoted by the Greek symbol: (eta). For most common UJTs, typical standard values range from 0.5 to 0.8.

If a small positive input voltage less than the voltage developed across the resistance, RB1 (VBB), is now applied to the Emitter input terminal, the diode p-n junction is reverse biased, resulting in a very high impedance and the device does not conduct. The UJT is turned "OFF," and no current flows.

The p-n junction becomes forward biased and the unijunction transistor begins to conduct when the Emitter input voltage exceeds VRB1 (or VBB + 0.7V, where 0.7V equals the p-n junction diode volt drop).

The Emitter current, IE, now flows from the Emitter into the Base region as a result. The resistive portion of the channel between the Emitter junction and the B1 terminal is reduced as a result of the additional Emitter current flowing into the Base. As the RB1 resistance is reduced to such a low value, the Emitter junction becomes even more forward biased, resulting in a larger current flow. As a result, at the Emitter terminal, there is a negative resistance.

Similarly, if the input voltage applied between the Emitter and B1 terminals falls below the breakdown voltage, the resistive value of RB1 rises to a high value.

The Unijunction Transistor can then be viewed as a voltage breakdown device. As a result, we can see that the resistance presented by RB1 is variable and depends on the value of the Emitter current, IE. Then, by forward biasing the Emitter junction in relation to B1, more current flows, lowering the resistance between the Emitter, E, and B1.

Since current flows into the UJT's Emitter, the resistive value of RB1 decreases and the voltage drop across it decreases, VRB1 must also decrease, allowing more current to flow and resulting in a negative resistance condition.

Applications of Unijunction Transistor

A unijunction transistor is most commonly used as a triggering device for SCRs and Triacs, but other UJT applications include saw-toothed generators, simple oscillators, phase control, and timing circuits.

The Relaxation Oscillator is the most basic of all UJT circuits, producing non-sinusoidal waveforms. The Emitter terminal of the unijunction transistor is connected to the junction of a series connected resistor and capacitor, RC circuit, in a basic and typical UJT relaxation oscillator circuit, as shown below.

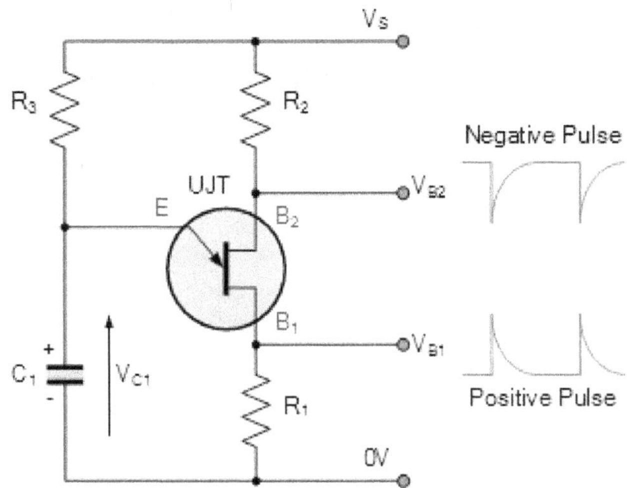

When a voltage (Vs) is applied for the first time, the unijunction transistor is "OFF," and the capacitor C1 is fully discharged but begins to charge exponentially via resistor R3. Because the UJT's Emitter is connected to the capacitor, when the charging voltage Vc across the capacitor exceeds the diode volt drop value, the p-n junction behaves like a normal diode and becomes forward biased, causing the UJT to conduct.

The unijunction transistor is turned "ON." At this point, the Emitter to B1 impedance collapses as the Emitter enters a low impedance saturated state, with Emitter current flowing through R1. The capacitor discharges rapidly through the UJT due to the low ohmic value of resistor R1, and a fast rising voltage pulse appears across R1.

Furthermore, because the capacitor discharges through the UJT faster than it charges up through resistor R3, the discharging time is much shorter than the charging time when the capacitor discharges through the low resistance UJT.

When the voltage across the capacitor falls below the p-n junction's holding point (VOFF), the UJT turns "OFF," and no current flows into the Emitter junction, so the capacitor charges up again through resistor R3, and the charging and discharging process between VON and VOFF is constantly repeated while a supply voltage, Vs, is applied.

Oscillator Waveforms of UJT

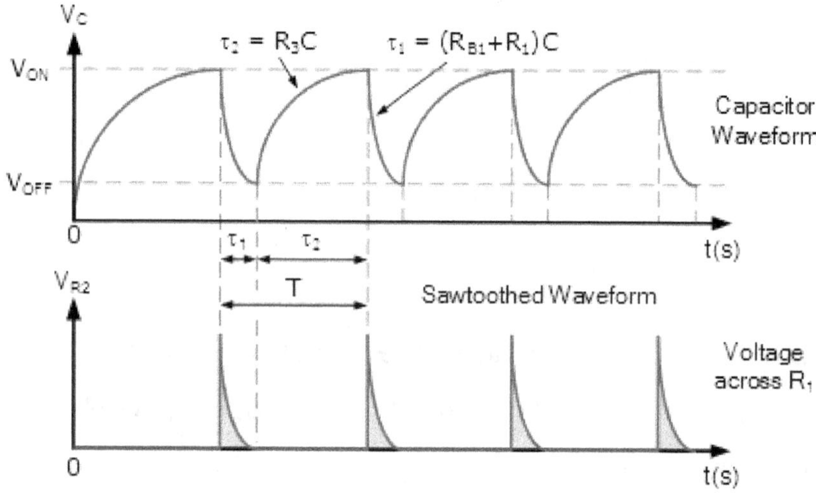

The unijunction oscillator then switches "ON" and "OFF" indefinitely without any feedback. The value of the charging resistance R3, in series with the capacitor C1 and the value of has a direct effect on the oscillator's frequency of operation.

The output pulse shape generated from the Base1 (B1) terminal is a sawtooth waveform, and changing the ohmic value of resistance, R3, which sets the RC time constant for charging the capacitor, is all that is required to regulate the time period. The charging time plus the capacitor's discharging time will be used to calculate the sawtoothed waveform's time period, T.

Since the discharge time (τ_1) is relatively shorter than the larger RC charging time (τ_2), oscillation's time period is more or less equivalent to $T \cong \tau_2$. Hence, oscillation's frequency is specified as $f = 1/T$.

Example of UJT Oscillator

The intrinsic stand-off ratio (η) for a 2N2646 Unijunction Transistor is 0.65, according to the data sheet. Calculate the timing resistor required to produce an oscillation frequency of 100Hz if a 100nF capacitor is used to generate the timing pulses.

1. The time frame is given as follows:

$$f = \frac{1}{T}, \quad \therefore T = \frac{1}{f} = \frac{1}{100} = 10\mathrm{mS}$$

2. Timing resistor's (R_3) value can be calculated as:

$$T = R_3 C \ln\left(\frac{1}{1\text{-}\eta}\right)$$

$$\therefore R_3 = \frac{T}{C \times \ln\left(\frac{1}{1\text{-}\eta}\right)} = \frac{10\mathrm{mS}}{100\mathrm{nF} \times \ln\left(\frac{1}{1\text{-}0.65}\right)}$$

$$\therefore R_3 = 95.238\Omega \quad \text{or} \quad 95.3\mathrm{k}\Omega$$

The charging resistor value required in this simple example is then calculated as 95.3k's to the nearest preferred value. Nevertheless, certain conditions must be met for the UJT relaxation oscillator to function properly, as the resistive value of R3 can be too large or too small.

For example, if R3 is too large (Megohms), the capacitor may not charge sufficiently to trigger the Unijunction's Emitter into conduction, but it must also be large enough to ensure that the UJT switches "OFF" once the capacitor has discharged to below the lower trigger voltage.

Similarly, if R3 is too low (a few hundred Ohms), the current flowing into the Emitter terminal may be sufficient to drive the device into its saturation region, preventing it from completely turning "OFF." The unijunction oscillator circuit would fail to oscillate in either case.

Speed Control Circuit of UJT

The above unijunction transistor circuit is commonly used to generate a series of pulses to fire and control a thyristor. We can adjust the speed of a universal AC or DC motor by using the UJT as a phase control triggering circuit in conjunction with an SCR or Triac, as shown.

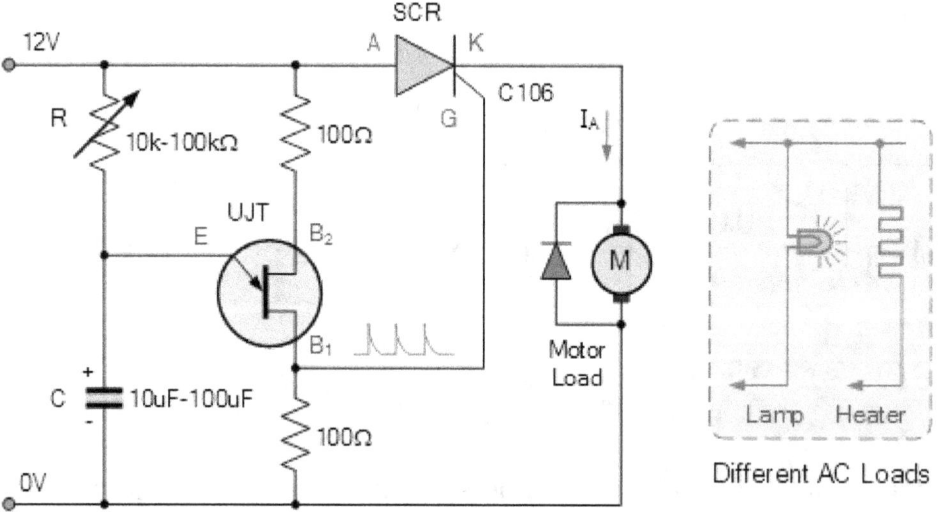

The speed of a universal series motor (or any other type of load, such as heaters or lamps) can be controlled by regulating the current flowing through the SCR using the circuit above. Simply change the frequency of the saw-tooth pulse, which is achieved by varying the value of the potentiometer, to control the motor's speed.

Unijunction Transistor in a brief

A Unijunction Transistor (UJT) is an electronic semiconductor device which contains only one p-n junction within an N-type (or P-type) lightly doped ohmic channel. The UJT has three terminals, one labeled Emitter (E) and two Bases (B1 and B2). The resistance between ohmic contacts B1 and B2 when the emitter is open circuited is known as the inter base resistance, R_{BB}. When measured with an ohmmeter, the static resistance of most common UJTs is typically between 4kΩ and 10kΩ.

The intrinsic stand-off ratio is the ratio of R_{B1} to R_{BB} and is denoted by the Greek symbol: η(eta). For most common UJTs, typical standard values range from 0.5 to 0.8. The unijunction transistor is a solid-state triggering device that can be used in a wide range of circuits and applications, from thyristor and triac firing to phase control circuits in saw-tooth generators. The UJT's negative resistance property makes it ideal for use as a simple relaxation oscillator.

It can oscillate independently without a tank circuit or complicated RC feedback network when connected as a relaxation oscillator. When connected in this manner, the unijunction transistor can generate a train of pulses of varying duration simply by varying the values of a single capacitor, (C), or resistor, (R) (R).

Unijunction transistors that are commonly available include the 2N1671, 2N2646, 2N2647, and others, with the 2N2646 being the most popular UJT for use in pulse and sawtooth generators and time delay circuits. Programmable UJTs are another type of unijunction transistor device that can have their switching parameters set by external resistors. The 2N6027 and 2N6028 are the most common Programmable Unijunction Transistors.

CHAPTER-7: SWITCH MODE POWER SUPPLY (SMPS)

For many years, linear voltage IC regulators have been the foundation of power supply designs because they are very good at supplying a continuous fixed voltage output. Linear voltage regulators are much more efficient and user-friendly than equivalent voltage regulator circuits made from discrete components such as a zener diode and a resistor, or transistors and even op-amps.

The 78... positive output voltage series and the 79... negative output voltage series are by far the most popular linear and fixed output voltage regulator types.

These two types of complementary voltage regulators generate a precise and stable voltage output ranging from about 5 volts to about 24 volts, which can be used in a variety of electronic circuits.

These three-terminal fixed voltage regulators are available in a variety of configurations, each with its own built-in voltage regulation and current limiting circuits. This enables us to design a wide range of power supply rails and outputs, either single or dual supply, suitable for the majority of electronic circuits and applications.

There are also variable voltage linear regulators that provide an output voltage that is continuously variable from just above zero to a few volts below its maximum voltage output.

A large and heavy step-down mains transformer, full-wave or half-wave diode rectification, and a filter circuit to remove any ripple content from the rectified DC to produce a suitably smooth DC output voltage are all common components of DC power supplies.

In addition, a voltage regulator or stabilizer circuit, either linear or switching, can be used to ensure that the power supply output voltage is properly regulated under varying load conditions. The following is an example of a typical DC power supply:

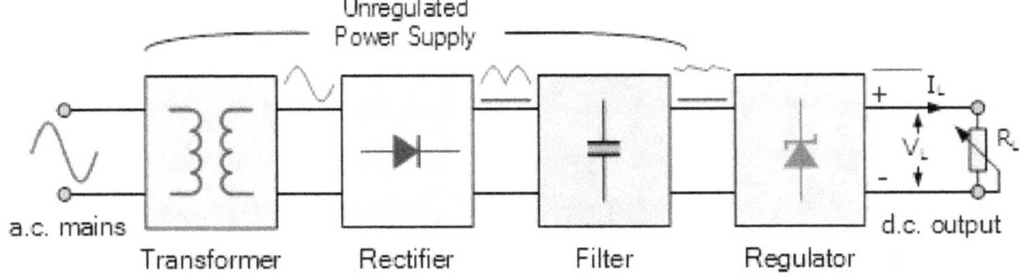

A large mains transformer (which also provides isolation between the input and output) and a series regulator circuit are common features in power supply designs. To produce the required output voltage, the regulator circuit could be made up of a single zener diode or a three-terminal linear series regulator.

A linear regulator has the advantage of only requiring an input capacitor, an output capacitor, and some feedback resistors to set the output voltage. Linear voltage regulators produce a regulated DC output by connecting an input and an output with a continuously conducting transistor and operating it in the linear region of its current-voltage (i-v) characteristics (hence the name).

As a result, the transistor behaves more like a variable resistance that adjusts to whatever value is required to maintain the desired output voltage. Consider the following circuit for a simple series pass transistor regulator:

Series Transistor Regulator Circuit

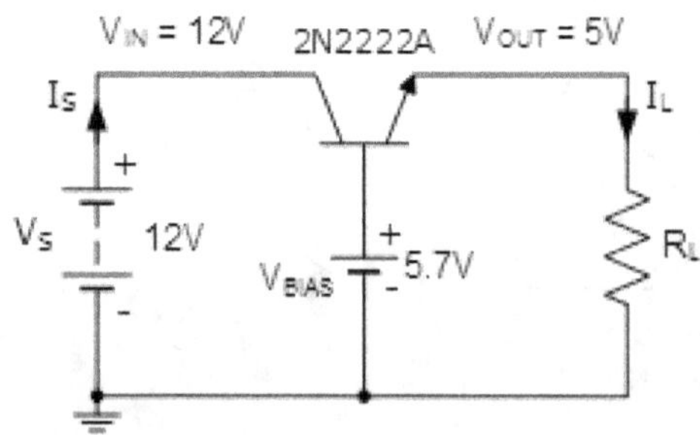

A single NPN transistor and a DC biasing voltage are used to set the required output voltage in this simple emitter-follower regulator circuit. Since an emitter follower circuit has unity voltage gain, a stable output can be obtained from the emitter terminal by applying a suitable biasing voltage to the transistor's base.

The output load current will be much higher than the base current because a transistor provides current gain, and even higher if a Darlington transistor arrangement is used.

The output voltage is controlled by the transistor's base voltage, which in this example is 5.7 volts to produce a 5 volt output to the load because approximately 0.7 volts is dropped across the transistor between the base and emitter terminals. Any value of emitter output voltage can then be obtained depending on the value of the base voltage. While this simple series regulator circuit will function, the series transistor will remain biased in its linear region, dissipating power in the form of heat. As all of the load current must pass through the series transistor, it has a low efficiency, wastes V*I power, and generates constant heat around it. Series voltage regulators also have the disadvantage of having a maximum continuous output current rating of only a few amperes or so, so they are typically used in applications requiring low power outputs.

When higher output voltage or current power demands are required, a switching regulator, also known as a switch-mode power supply, is typically used to convert the mains voltage into the desired higher power output.

Switch Mode Power Supplies (SMPS) are becoming more common and have largely replaced traditional linear AC-to-DC power supplies as a means of reducing power consumption, heat dissipation, size, and weight. As switch-mode power supplies become a much more mature technology, they can now be found in most PCs, power amplifiers, TVs, dc motor drives, and just about anything that requires a highly efficient supply.

A switch mode power supply (SMPS) is a type of power supply that uses semiconductor switching techniques to provide the required output voltage rather than standard linear methods. A power switching stage and a control circuit comprise the basic switching converter. The power switching stage converts power from the circuit's input voltage, VIN, to its output voltage, VOUT, including output filtering.

The main advantage of switch mode power supplies over standard linear regulators is their higher efficiency, which is achieved by internally switching a transistor (or power MOSFET) between its "ON" (saturated) and "OFF" (cut-off) states, both of which produce lower power dissipation. This means that when the switching transistor is fully "ON" and conducting current, the voltage drop across it is at its lowest, and when the transistor is completely "OFF," no current flows through it. As a result, the transistor functions as an ideal ON/OFF switch.

In contrast to linear regulators, which only provide step-down voltage regulation, a switch mode power supply can provide step-down, step-up, and negation of the input voltage by employing one or more of the three basic switch mode circuit topologies: Buck, Boost, and Buck-Boost. Within the basic SMPS circuit, these names refer to how the transistor switch, inductor, and smoothing capacitor are connected.

Buck Switch Mode Power Supply

The Buck switching regulator is a switch mode power supply circuit that is designed to efficiently reduce DC voltage from a higher to a lower voltage, that is, it subtracts or "Bucks" the supply voltage, reducing the voltage available at the output terminals without changing the polarity.

In other words, the buck switching regulator is a step-down regulator circuit, so a buck converter, for example, can convert +12 volts to +5 volts. The buck switching regulator is a DC-to-DC converter that is one of the most basic and widely used types of switching regulators.

The buck switching regulator utilizes a series transistor or power MOSFET (ideally an insulated gate bipolar transistor, or IGBT) as its main switching device while using in a switch mode power supply configuration, as shown below.

The Buck Switching Regulator

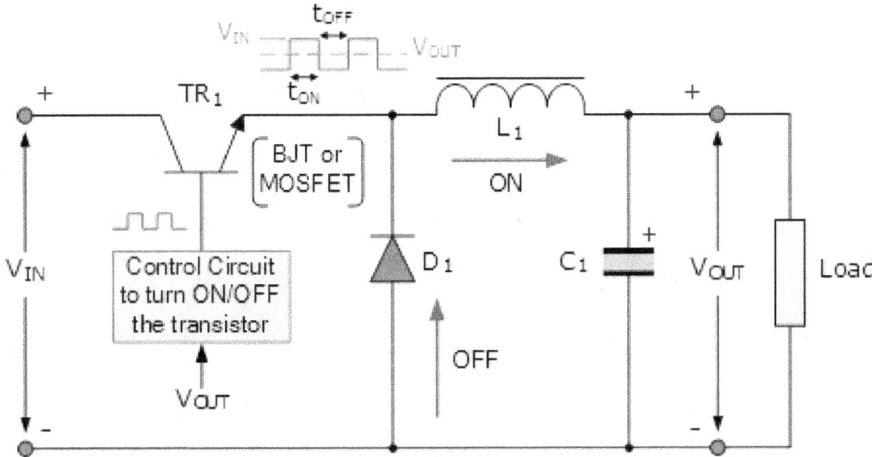

A series transistor switch, TR1, with an associated drive circuit that keeps the output voltage as close to the desired level as possible, a diode, D1, an inductor, L1, and a smoothing capacitor, C1, make up the basic circuit configuration for a buck converter.

Depending on whether the switching transistor TR1 is turned "ON" or "OFF," the buck converter can operate in one of two modes. When the transistor is biased "ON" (switch closed), diode D1 becomes reverse biased, causing current to flow through the inductor to the connected load at the output, charging the capacitor, C1.

According to Faraday's law, as a changing current flows through the inductor coil, it creates a back-emf that opposes the flow of current until it reaches a steady state, creating a magnetic field around the inductor, L1.

As long as TR1 is closed, this situation will continue indefinitely. The input voltage is immediately disconnected from the emitter circuit when transistor TR1 is turned "OFF" (switch open) by the controlling circuitry, causing the magnetic field around the inductor to collapse, inducing a reverse voltage across the inductor. The diode becomes forward biased as a result of the reverse voltage, and the stored energy in the inductors magnetic field forces current to flow in the same direction through the load and back through the diode.

The inductor, L1, then acts as a source, supplying current to the load until all of the inductor's stored energy is returned to the circuit, or until the transistor switch closes again, whichever comes first. The capacitor discharges at the same time, supplying current to the load. The inductor and capacitor work together to create an LC filter, which smooths out any ripple caused by the transistor's switching action. As a result, current is supplied from the supply when the transistor solid state switch is closed, and current is supplied by the inductor when the transistor solid state switch is open.

The current flowing through the inductor is always in the same direction, whether it comes directly from the supply or through the diode, but at different times during the switching cycle. The average output voltage value will be related to the duty cycle, D, which is defined as the conduction time of the transistor switch during one full switching cycle. If VIN is the supply voltage and tON and tOFF are the transistor switch's "ON" and "OFF" times, then VOUT is the output voltage.

Buck Converter Duty Cycle

$$V_{OUT} = \frac{t_{ON}}{(t_{ON} + t_{OFF})} \times V_{IN}$$

The buck converters duty cycle can also be defined as:

$$D = \frac{t_{ON}}{(t_{ON} + t_{OFF})} = \frac{t_{ON}}{Total\ Time} = \frac{t_{ON}}{T}$$

$$\therefore D \approx \frac{V_{OUT}}{V_{IN}} \quad or \quad V_{OUT} = D\,V_{IN}$$

As a result, the higher the duty cycle, the higher the switch mode power supply's average DC output voltage. As the duty cycle, D, can never reach one (unity), the output voltage will always be lower than the input voltage, resulting in a step-down voltage regulator.

Voltage regulation is achieved by varying the duty cycle, and with high switching speeds of up to 200kHz, smaller components can be used, resulting in a significant reduction in the size and weight of a switch mode power supply.

The inductor-capacitor (LC) arrangement of the buck converter also provides excellent inductor current filtering. The buck converter should ideally be switched on continuously so that the inductor current never drops to zero.

The ideal buck converter could have efficiencies as high as 100 percent with ideal components, such as zero voltage drop and switching losses in the "ON" state. The Boost Converter is another operation of the fundamental switching regulator that acts as a step-up voltage regulator, in addition to the step-down buck switching regulator for the basic design of a switch mode power supply.

Boost Switch Mode Power Supply

Another form of switch mode power supply circuit is the boost switching regulator. It contains the same components as the previous buck converter, however they are arranged differently this time. The boost converter adds to or "Boosts" the supply voltage, so increasing the available voltage at the output terminals without affecting the polarity.

In other terms, the boost switching regulator is a step-up regulator circuit, allowing a boost converter to convert +5 volts to +12 volts, for example. A series switching transistor is used in the fundamental construction of the buck switching regulator, as we saw previously.

The boost switching regulator differs in that it controls the output voltage from the switch mode power supply using a parallel linked switching transistor. Electrical energy only goes through the inductor to the load when the transistor is biased "OFF" (switch open) as depicted, because the transistor switch is essentially linked in parallel with the output.

The Boost Switching Regulator

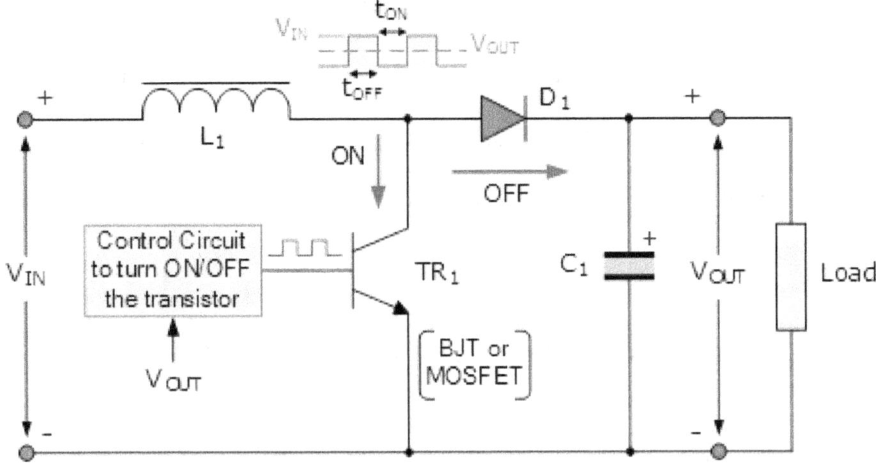

When the transistor switch is fully on in the Boost Converter circuit, electrical energy from the supply, VIN, flows through the inductor and transistor switch before returning to the supply. As a result, since the saturated transistor switch essentially produces a short-circuit to the output, none of it reaches the output. Because the inductor's inner path to the supply is shorter, the current flowing through it is increased.

Meanwhile, as the capacitor discharges through the load, diode D1 becomes reverse biased as its anode is connected to ground via the transistor switch, and the voltage level on the output remains fairly constant.

The input supply is now connected to the output via the series connected inductor and diode when the transistor is fully turned off. The induced energy held in the inductor is pushed to the output by V_{IN}, through the now forward biased diode, as the inductor field declines. As a result of all of this, the induced voltage across the inductor L1 reverses and adds to the voltage of the input supply, resulting in $V_{IN} + V_L$ as the total output voltage.

The input supply via the diode now returns current to the smoothing capacitor, C1, which was utilized to power the load while the transistor switch was closed. The diode current is then provided to the capacitor, which is always "ON" or "OFF" since the diode is constantly cycled between forward and reverse states by the transistor switching operation.

The smoothing capacitor must then be large enough to generate a smooth and consistent output. The negative induced voltage across the inductor L1 adds to the source voltage, V_{IN}, driving the inductor current into the load. The steady state output voltage of a boost converter is given by:

$$V_{OUT} = V_{IN}\frac{1}{(1 - \text{duty cycle})} = V_{IN}\left(\frac{1}{1 - D}\right)$$

The output voltage of the boost converter, like that of the prior buck converter, is determined by the input voltage and duty cycle. As a result, output regulation is performed by adjusting the duty cycle. Not to mention that the inductor value, load current, and output capacitor are all unaffected by this equation.

Depending on whether we need a step-down (buck) or step-up (boost) output voltage, the basic functionality of a non-isolated switch mode power supply circuit can use either a buck converter or a boost converter architecture.

When buck converters are the more frequent SMPS switching arrangement, boost converters are employed in capacitive circuit applications like battery chargers, photo-flashes, strobe flashes, and so on, because the capacitor delivers all of the load current while the switch is closed. However, these two basic switching topologies can be combined into a single non-isolating switching regulator circuit known as a Buck-Boost Converter.

Buck-Boost Switching Regulator

Buck-Boost switching regulators combine buck and boost converters to create an inverted (negative) output voltage that can be larger or less than the input voltage depending on the duty cycle.

The buck-boost converter is a boost converter variant in which the inverting converter only sends the energy stored in the inductor, L1, to the load. The circuit for a basic buck-boost switch mode power supply is shown below.

The Buck-Boost Switching Regulator

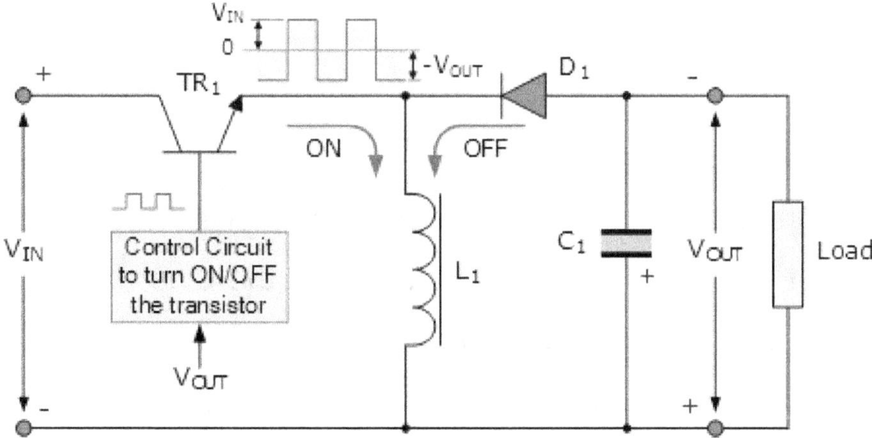

The voltage across the inductor is equal to the supply voltage when the transistor switch, TR1, is fully on (closed), hence the inductor stores energy from the input supply. Because the output diode, D1, is reverse biased, no current is transmitted to the associated load. The diode becomes forward biased when the transistor switch is fully off (open), and the energy previously stored in the inductor is delivered to the load.

When the switch is "ON," the DC supply delivers energy into the inductor (via the switch), but none to the output; when the switch is "OFF," the voltage across the inductor reverses as the inductor now becomes a source of energy, and the energy previously stored in the inductor is switched to the output (via the diode), and none from the input DC source. When the switching transistor is "OFF," the voltage decreased across the load equals the inductor voltage.

As a result, depending on the duty cycle, the magnitude of the inverted output voltage can be higher or smaller (or equal to) the magnitude of the input voltage. A positive-to-negative buck-boost converter, for instance, can convert 5 volts to 12 volts (step-up) or 12 volts to 5 volts (step-down) (step-down). The steady state output voltage, VOUT, of a buck-boost switching regulator is given as:

$$V_{OUT} = V_{IN}\left(\frac{D}{1-D}\right)$$

The buck-boost regulator derives its name from the fact that it generates an output voltage that can be greater (like a boost power stage) or lower (like a buck power stage) than the input voltage. The output voltage, on the other hand, is polarized in the opposite direction as the input voltage.

Switch Mode Power Supply in a brief

The typical switch mode power supply, or SMPS, converts an unregulated DC input voltage to a regulated and smooth DC output voltage at various voltage levels using solid-state switches. A genuine DC voltage from a battery or solar panel, or a rectified DC voltage from an AC source using a diode bridge and some further capacitive filtering, can be used as the input supply.

The power transistor, MOSFET or IGFET, is used in many power control applications in its switching mode, where it is repeatedly turned "ON" and "OFF" at high speed. Because the transistor is either fully on and conducting (saturated) or entirely off (full-off), the power efficiency of the regulator can be fairly high (cut-off).

There are various types of DC-to-DC converter (as opposed to a DC-to-AC converter, which is an inverter) configurations available, with the Buck, Boost, and Buck-Boost switching regulators being the three primary switching power supply topologies examined here. All three topologies are non-isolated, meaning their input and output voltages are connected to a single ground line. The steady-state duty cycles, relationship between the input and output current, and output voltage ripple produced by the solid-state switch action are all unique to each switching regulator design. The frequency responsiveness of the switching action to the output voltage is another significant aspect of these switch mode power supply topologies.

The percentage control of the time the switching transistor is in the "ON" state compared to the entire ON/OFF period is used to regulate the output voltage. This ratio is known as the duty cycle, and it may be changed by adjusting the duty cycle (D the magnitude of the output voltage).

Within the switch mode power supply design, the use of a single inductor and diode, as well as rapid switching solid-state switches capable of working at switching frequencies in the kilohertz region, considerably reduces the size and weight of the power supply. This is due to the lack of huge, heavy step-down (or step-up) voltage mains transformers in their design. If electrical isolation between the input and output terminals is necessary, a transformer must be included before the converter.

The buck (subtractive) and boost (additive) converters are the two most common non-isolated switching designs. A buck converter is a switch-mode power supply that is used to transfer electrical energy from one voltage to a lower voltage. A series-connected switching transistor powers the buck converter. Due to the obvious duty cycle, D< 1, the buck's output voltage is always less than the input voltage, V_{IN}.

A boost converter is a switch-mode power supply that converts electrical energy from one voltage to a higher voltage. The boost converter uses a parallel connected switching transistor to create a direct current route between V_{IN} and VOUT through the inductor L1 and diode D1. This means the output has no protection against short-circuits. The output voltage of a boost converter may be regulated by altering the duty cycle, (D), and with D < 1, the DC output from the boost converter is greater than the input voltage V_{IN} due to the inductors self-induced voltage.

CHAPTER-8: TRANSIENT SUPPRESSION DEVICES

The amount of energy produced as a result of over voltage spikes and surges can be considerably reduced by transient suppression devices. We'd like to believe that the AC or DC power supplies we utilize to run our circuits are clean and well-regulated.

Switching AC inductive loads or DC relay contacts and DC motors as part of a microcontroller project, on the other hand, all combine to provide a power supply that is difficult to maintain. When an inductive or reactive load, such as a motor, a solenoid coil, or a relay coil, is rapidly switched off, inductive switching transients occur.

Transients are pretty steep voltage steps that occur in electrical circuits as a result of the sudden release of previously stored energy, whether inductive or capacitive, resulting in a high voltage transient, or surge. A transient voltage spike in the form of a steep impulse of energy is created when energy is suddenly released back into the circuit due to some switching action. This energy can theoretically be of any infinite magnitude.

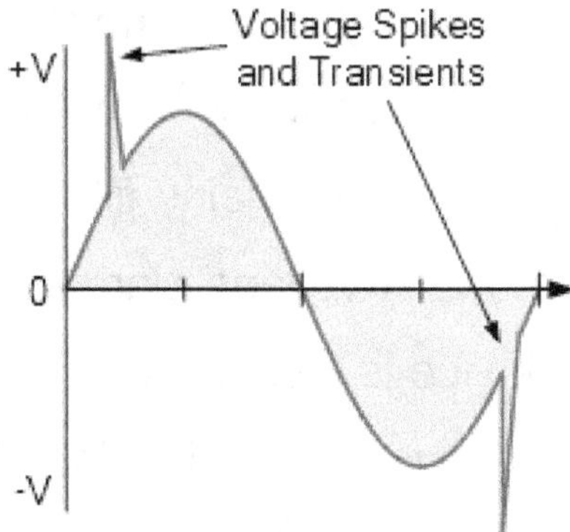

This high dv/dt transient switching spike might last for a very short time (milli-seconds or micro-seconds), or it can happen infrequently over short periods of time, such as two or three times a day. It's also important to keep in mind that voltage transients don't always begin at zero volts or at the start of a cycle, but can be overlaid on another voltage level.

Transients need to be repressed and controlled, because they can destroy electronic equipment any time. Arc contacts, filters, and solid state semiconductor devices are all examples of transient suppression devices.

Discrete semiconductor transient suppression devices like the Metal-oxide Varistor, or MOV, are by far the most prevalent because they come in a wide range of energy absorption and voltage ratings, allowing tight control over unwanted and potentially damaging transients or voltage spikes.

Transient suppression devices can be used in series with the load to either attenuate or reduce the energy value of a transient, preventing it from propagating through the circuit, or in parallel with the load to divert the transient away from the circuit, usually to ground, and thus limit or clamp the residual voltage.

Low-pass filters connected in series with the load circuit are commonly used to reduce voltage transients. Since a voltage transient is usually a fast-moving, high-frequency spike, the filter attenuates or blocks the high-frequency transient while leaving the low-frequency power or signal component alone. Mains filtered extension cords are a nice example of transient attenuators. A voltage-clamping type device or what is widely known as a crowbar type device is typically used to divert a transient. Because the current flowing through these parallel linked devices is not linear to the voltage across their terminals as defined by Ohms Law, they have a nonlinear impedance characteristic.

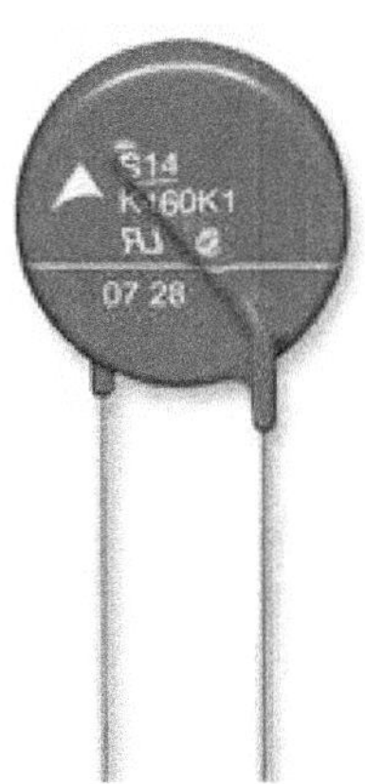

The impedance of a voltage-clamping device, like MOV, varies depending on the current flowing through it or the voltage across its terminal. The gadget has a high impedance under normal steady-state operation conditions and hence has no influence on the connected circuit.

When a voltage transient occurs, however, the device's impedance changes, increasing the current pulled through it as the voltage across it increases. As a result, the transient voltage appears to be clamped. Since the huge increase in current causes the device to dissipate a lot of energy, the volt-ampere characteristic of clamping devices is usually time-dependent.

Crowbar devices are a kind of transient suppression device that uses a switching type turn-on operation to divert excess voltage spikes away from a circuit. Crowbars work similarly to zener diodes in that they have no influence on the circuit under normal steady-state conditions. They quickly switch "ON" when a transient is detected, providing an extremely low impedance channel that diverts the transient away from the parallel-connected load. Then, based on the kind of connection and operation, discrete transient suppression devices can be classified into three basic groups.

> Low Pass Filters linked in series (blocking).
> Voltage Clampers and Clippers connected in parallel (shunting).

➢ Connected Crowbar devices in parallel (shunting).

This can be demonstrated using the following example:

Transient Suppression Devices

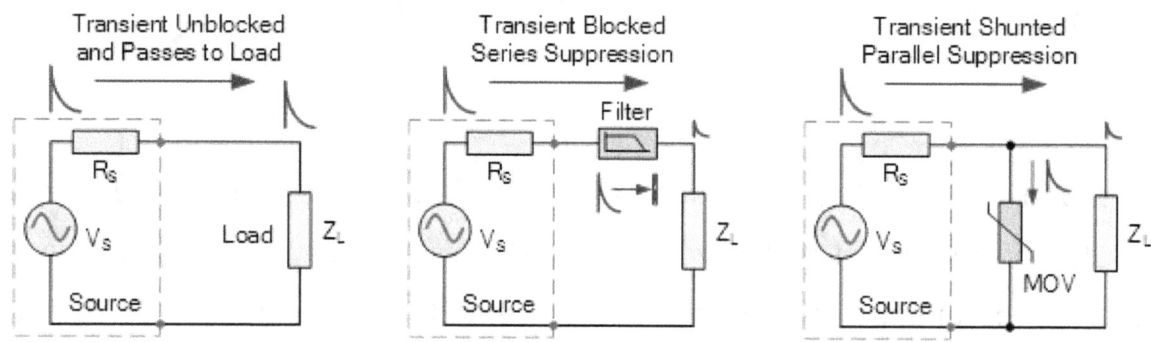

Series Transient Suppression Filters

An AC power line's transients can range from a few volts to many kilovolts above the regular mains voltage. Filter circuits effectively reduce these mains born transients by inserting a 100Hz filter in series with the linked load in suppression devices that attenuate or block these transients.

A fast switching voltage transient's frequency component can be significantly higher than the AC source's slow moving fundamental frequency. As a result, using a low-pass filter section between the source and the load is a natural choice for attenuating and controlling these undesired transients.

Low pass filters, like an LC filter, can reduce high frequency transients while allowing low-frequency power or signal to pass through unaffected. A resistor-capacitor RC filter put directly across the power line to attenuate any high-frequency transients is the simplest form of transient suppression filter. Inductances and capacitors are used to create multistage LC filters for AC power applications, and the degree of attenuation is determined by the number of LC stages in the filter. The diagram below depicts a typical series-connected AC mains transient suppression filter.

Typical Transient Suppression Filter

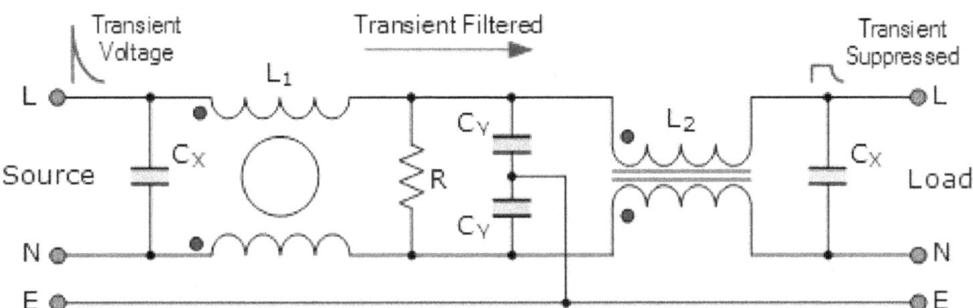

Throughout the frequency range, this basic two-stage low-pass AC filter provides a large insertion loss between line-to-line and line-to-ground, providing effective transient voltage protection by preventing any high-frequency transient and noise from reaching the connected load equipment. These mains power filters can also help reduce any radio-frequency interference or emissions emitted by the power supply, in addition to decreasing voltage spikes and transients.

Voltage Clamping Transient Suppressors

The amplitude of a transient across a circuit is limited by voltage clampers. When a predefined threshold voltage is surpassed, a voltage clamping device begins to conduct, then switches back to non-conducting mode when the overvoltage falls below the threshold level. As a result, the clamping mechanism reduces overvoltage spikes to a safe level.

To protect the load from undesirable high dv/dt voltage transients, voltage clamping devices are often installed across the supply and in parallel with the load. A voltage clamper can be as simple as a zener diode across a DC supply, but we need to employ metal oxide varistor (MOV), suppression diodes, or voltage dependent resistors (VDR) for bi-directional AC supplies.

Since voltage clamping devices divert surge currents rather than absorbing them like a filter, it's important to make sure the channel utilized to divert the transient doesn't cause or exacerbate the circuit's difficulties.

Zener Diode Transient Suppressors

Zener diodes are utilized for unidirectional DC supply protection because they function like conventional diodes in the forward biased direction but breakdown and conduct in the reverse biased direction. The reverse breakdown voltage of a zener diode, VZ, can thus be utilized as a reference or clamping voltage level.

VZ zener diodes have a high impedance to the supply and conduct very little leakage current when operated in the reverse direction and below their zener breakdown voltage. When the voltage across the zener exceeds its zener voltage, it begins to breakdown, with conduction progressively increasing as the voltage across it rises, resulting in a very low impedance path to the over voltage transient.

Zener Transient Suppression

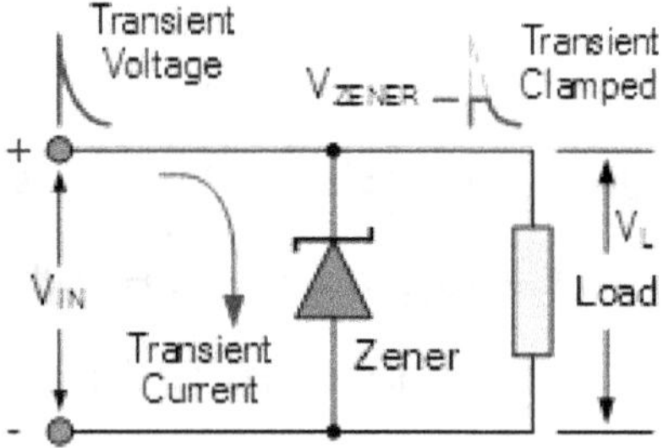

The zener diode is practically "invisible" until a transient voltage appears when connected across a supply or across the components being protected because it has a high impedance below its reverse breakdown voltage and a low impedance above its reverse breakdown voltage.

While the zener is in breakdown mode, that is, when suppressing a transient, the diode clamps the over voltage instantaneously to limit the spike to a safe level and then returns to normal operation after the transient voltage falls below the zener voltage, VZ.

The clamping voltage, VC, is thus equal to the reverse breakdown voltage of the zener. Because of its clamping properties, the zener diode is employed to decrease transients by clamping potentially hazardous currents away from the protected load.

The zener diode's surge current and power capabilities are roughly proportional to its junction area. The majority of zener diodes are made to function at low current and voltage levels. Avalanche Diodes are zener diodes that are engineered to function at greater voltage levels and absorb bigger surge currents without causing damage.

Because of their forward biased diode characteristics, a single zener diode can only be utilized for transient suppression on steady state DC supply, as we previously said. We can leverage the clamping properties of two zener diodes by connecting them "back-to-back" across a bidirectional AC supply.

Zener Transient Suppression

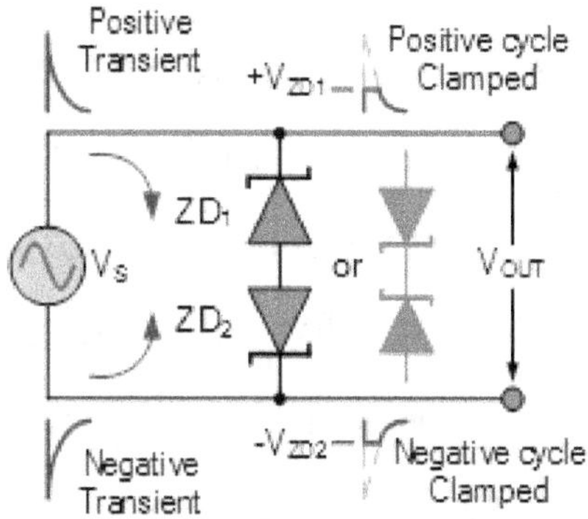

We can now safeguard the positive half cycle against overvoltage transients with one zener diode and the negative half cycle with the other by connecting two zener diodes back-to-back. If the reverse breakdown voltages of both zener diodes are the same, a transient voltage of either polarity will be clamped at the same zener voltage level because one zener diode will be effectively in reverse bias mode while the other will be in forward bias mode.

While two back-to-back zener diodes can be used to suppress AC transients, transient voltage suppressor (TVS) devices have opposing junctions incorporated into one device, making them excellent for AC power applications. A variety of voltage and power levels are available in bidirectional avalanche diodes.

MOV Transient Suppressors

While zener diodes and rapid recovery avalanche diodes are quick acting and good at clamping overvoltages, metal oxide varistors, or MOVs, are the most used overvoltage suppression clamping technology. Metal oxide varistors, in addition to their high voltage ratings, can handle significantly bigger surge currents, albeit at a slower rate, and can be used in both DC and AC power lines to protect against voltage extremes like overvoltage transients.

The MOV is a semiconductor voltage-dependent variable resistor that is connected in parallel (shunt) with the load or component that has to be safeguarded. MOVs have a high resistance at low voltage and a low resistance at high voltage, and their nonlinear voltage-current properties make them effective in preventing power-line surges and overvoltage transients. MOVs act similarly to back-to-back zener diodes in that they can be used for bidirectional voltage clamping, with transient conduction increasing as the voltage across it increases. Since these compact disk-shaped metal-oxide varistors have high breakdown voltages in both directions and can absorb more energy, they are frequently rated in joules rather than watts.

MOV Transient Suppression

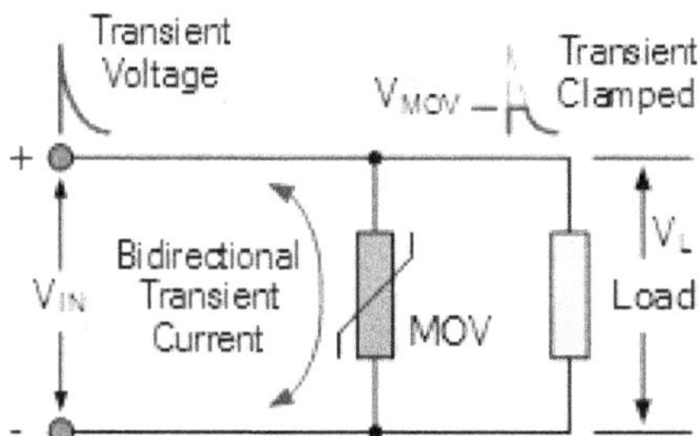

Metal oxide varistors, as voltage clamping devices, offer extremely high resistances when the voltage between their terminals is less than their pre-determined breakdown value, acting more like a voltage dependant resistor (VDR). The electrical characteristics of the device alter when exposed to high transient voltage of either polarity, and its resistance becomes very small, clamping the voltage to a safe level. When employed as a transient suppression device, the fundamental objective of the metal oxide varistor is to clamp the voltage appearing across it to a safe level, because in most applications, the device is connected in parallel with the circuit or device to be protected.

Crowbar Transient Suppressors

Crowbar protection is a sort of parallel (shunt) coupled transient suppression device. When a predetermined threshold voltage is exceeded, electronic crowbar devices conduct by triggering to a conductive on-state, resulting in a voltage drop of only a few volts, hence the name crowbar.

When a trigger voltage is reached, crowbar devices and circuits effectively create a short circuit. They are commonly found in stabilized power supplies that have been designed to produce a fixed output voltage, such as a constant 12 volts or 5 volts, but they can also be used to protect a circuit or load from transient over voltages.

Active crowbar circuits based on semiconductors are arranged in parallel (shunt) with the load and are capable of attenuating very significant surge currents. Because thyristors have a low "on-state" voltage and can keep voltage levels well below dangerous levels, they are commonly employed in crowbar circuits.

As they serve as a very low impedance type switch, they can divert a significant amount of transient energy to ground once fired. Because the power supply is shorted by the crowbar device, the output voltage will be zero, this short circuit may cause circuit fuses or circuit breakers to operate if additional commutation circuitry is not provided to turn "OFF" the crowbar clamp once switched "ON," especially in a DC system. Consider the circuit for clamping a crowbar shown below.

Basic Crowbar Clamping Circuit

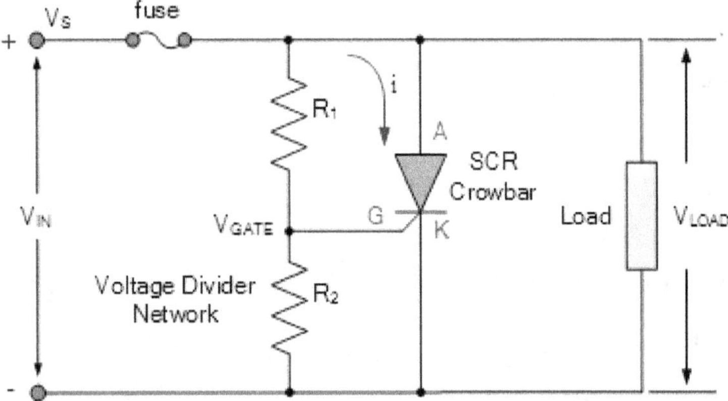

A thyristor or SCR is connected between the supply and the load, with the voltage divider circuit set up by resistors R1 and R2 to bias the thyristor's gate at a low enough level that it does not turn on during normal operation. The SCR is then turned off and becomes non-conducting.

When an over voltage transient occurs and exceeds a specified level, the voltage drop across resistor R2 climbs to the point where the gate of the SCR is triggered into conduction, clamping the voltage transient and safeguarding the load. The issue is that while the load is protected from over voltage, the power supply is not, resulting in the power supply's fuse blowing.

The load's protection against the transient caused by short-circuiting the power supply may then be larger than the triggering event. Triacs can be employed as a crowbar device and triggered into conduction in the same way as thyristors for excess voltage protection of AC power sources.

The advantage of utilizing thyristors or triacs for AC supply crowbar protection is that they turn off automatically every half-cycle. If the crowbar device is triggered by a short-duration transient of a fraction of a millisecond, the shunting action will only short-circuit the AC power line to which it is attached for at least one half-cycle, which may be too rapid for the fuse-link to blow.

Zener Crowbar Transient Suppressors

By introducing a zener diode to detect an overvoltage state, we may increase the transient sensing and performance of the simple crowbar circuit above. The resistive voltage divider circuit has been replaced with a zener diode in this example.

Zener Crowbar Clamping Circuit

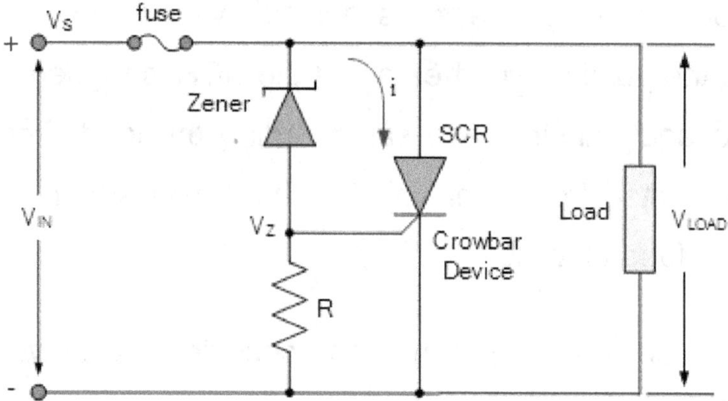

The zener diode, which acts as a transient detection component and whose zener voltage, VZ rating determines the voltage level at which the SCR turns on, monitors the DC supply voltage, VS.

When the DC supply voltage is less than the zener diode's reverse bias rating, the zener diode does not conduct, and no voltage or current is applied to the SCR's gate, leaving the SCR turned "OFF" and non-conducting.

When the supply voltage exceeds the zener voltage rating, such as during an overvoltage transient, the zener diode begins to conduct, allowing gate current to flow into the SCR, turning it "ON" and shorting out the load supply voltage and blowing the fuse. The load is then protected from transient voltages above the zener voltage, VZ, because the zener diode only carries the gate current for the SCR to turn "ON," as the SCR carries the majority of the shunt current.

Since this zener crowbar circuit develops on the basic voltage divider network, it has a soft turn-on action due to the curved knee at zener breakdown voltage rather than a sharp rise. By adding some voltage gain to the detection and triggering circuit in the form of a single amplifier circuit or op-amp circuit, the basic crowbar circuit can be further customized and developed.

Thyristors with built-in overvoltage triggers have been designed to crowbar unidirectional or bidirectional transients and voltage surges. Like the RCA SK9345 series of IC crowbars, that are designed to protect 15 volt power supplies, the SK9346 series, which protects 112 volt supplies, and the SK9347 series, which protects 115 volt supplies.

An integrated circuit with a built-in zener diode, transistors, and an SCR are used in all of them. The MC3423 over voltage crowbar sensing circuit is a single IC that works with an external crowbar SCR to sense voltage.

Transient Suppression Devices in a brief

We are becoming more reliant on over voltage protective devices to protect our equipment from voltage spikes and surges as we use more electronic devices in our daily lives. Inductive or capacitive switching circuits that release sudden, high-voltage spikes are the most common cause of transient over voltages. These voltage spikes and surges are superimposed on top of a steady-state value, such as an AC mains waveform, and can contain high energy for a short period of time or intermittently for short periods of time.

Over voltage protection circuits come in a variety of shapes and sizes, ranging from series connected filters that pass power-line frequency voltages and currents while denying unwanted high frequency harmonics and noise to parallel connected clamping and crowbar circuits that dissipate the over voltage to ground.

A capacitor put across the voltage source is the most basic type of AC power-line filter. The capacitor's impedance changes, causing high-frequency transients to be attenuated. In most cases, the transient suppression device is connected in series with the protected load or with a component that needs to be protected.

A voltage suppression circuit's main goal is to keep the voltage at a safe level. Metal oxide varistors, MOVs, and Zener Diodes are the most common voltage clamping devices. MOVs are best for bidirectional AC power supply protection, while zener diodes are best for low-energy DC supplies.

Solid-state crowbar circuits that use an SCR or triac as a "crowbar" quickly short the voltage transient across the power supply, causing the over-voltage fuse to blow. Many different combinations are possible with hybrid transient/surge protectors, which incorporate a crowbar with a clamp, or a clamp/crowbar with a filter, in one module.

CHAPTER-9: SOLID STATE RELAY

Solid State Relays are semiconductor similar of electromechanical relays which can be used to regulate electrical loads without using moving parts. Unlike electro-mechanical relays (EMR), which use coils, magnetic fields, springs, and mechanical contacts to operate and switch a supply, the solid state relay (SSR) has no moving parts and instead uses the electrical and optical properties of solid state semiconductors to perform input to output isolation and switching functions.

SSRs, like any other electromechanical relay, help in providing complete electrical isolation between their input and output contacts, with the output acting like a conventional electrical switch in that it has a very high, almost infinite resistance during no conduction (open) and a very low resistance during conduction (closed).

Instead of the usual mechanical normally-open (NO) contacts, solid state relays can be designed to switch both AC and DC currents by using an SCR, TRIAC, or switching transistor output. While both solid state and electro-mechanical relays have a low voltage input that is electrically disconnected from the output which switches and regulates a load, electro-mechanical relays, especially large power relays and contactors, have a limited contact life cycle, take up a lot of space, and have slower switch speeds. There are no such limitations with solid state relays.

The key benefits of solid state relays over traditional electro-mechanical relays are that they have no movable parts to carry out, and thus no contact bounce issues, and that they can switch both "ON" and "OFF" much faster than a mechanical relay's armature can move, as well as zero voltage and zero current turn-on and turn-off, which eliminates electrical noise and transients.

Solid state relays are available in a variety of standard off-the-shelf packages with output switching capabilities ranging from a few volts or amperes to many hundreds of volts and amperes. Solid state relays with very high current ratings (150A or more) require power semiconductor and heat dipping which is very costly. Therefore, cheaper electro-mechanical contactors are still used.

A small input voltage, typically 3 to 32 volts DC, can be used to control a much larger output voltage or current, similar to an electro-mechanical relay. 240V, 10Amps, for example. This makes opto-isolators ideal for microcontroller, PIC, and Arduino interfacing because a low-current, 5-volt signal from, say, a microcontroller or logic gate, can be used to control a specific circuit load.

Input of Solid State Relay

An opto-isolator (also known as an optocoupler) is a main component of a solid state relay (SSR) that contains one (or more) infrared light-emitting diode (LED) light source and a photosensitive device in a single case.

The opto-isolator separates the input and output. The LED light source is connected to the SSR's input drive section and optically couples to an adjacent photosensitive transistor, darlington pair, or triac via a gap. When a current is passed through the LED, it illuminates and directs its light across the gap to a phototransistor/phototriac.

In this way, by energizing this LED with a low-voltage signal, the output of an opto-coupled SSR is turned "ON." Since the only connection between the input and output is a beam of light, this internal opto-isolation achieves high voltage isolation (typically several thousand volts). The opto-isolator not only provides greater input/output isolation, but it can also transmit dc and low-frequency signals.

Furthermore, the LED and photo-sensitive device could be completely separate and optically coupled via an optical fiber. An SSR's input circuitry can be as simple as a single current limiting resistor in series with the opto-LED, isolator's or as complex as a circuit with rectification, current regulation, reverse polarity protection, filtering, and so on.

A voltage higher than its threshold voltage (usually 3 volts DC) must be applied to the input terminals of a solid state relay to activate or turn it "ON" (similar to the electro-mechanical relay coil). As shown, this DC signal can be generated by a mechanical switch, a logic gate, or a microcontroller.

DC Input Circuit of Solid State Relay

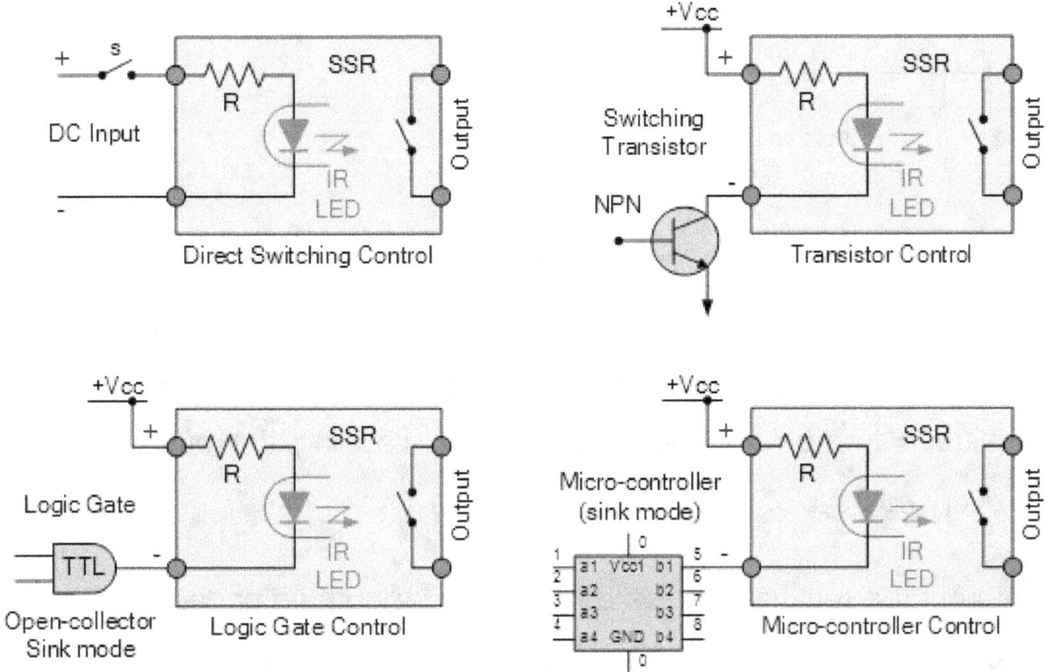

During using mechanical contacts, switches, pushbuttons, and other relay contacts as the activating signal, the supply voltage can be equal to the SSR's minimum input voltage value, whereas when using solid state devices like transistors, gates, and microcontrollers, the minimum supply voltage must be one or two volts above the SSR's turn-on voltage to account for the switching devices internal voltage drop.

We can also use a sinusoidal waveform by adding a bridge rectifier for full-wave rectification and a filter circuit to the DC input, in addition to using a DC voltage, either sinking or sourcing, to switch the solid state relay into conduction.

AC Input Circuit of Solid State Relay

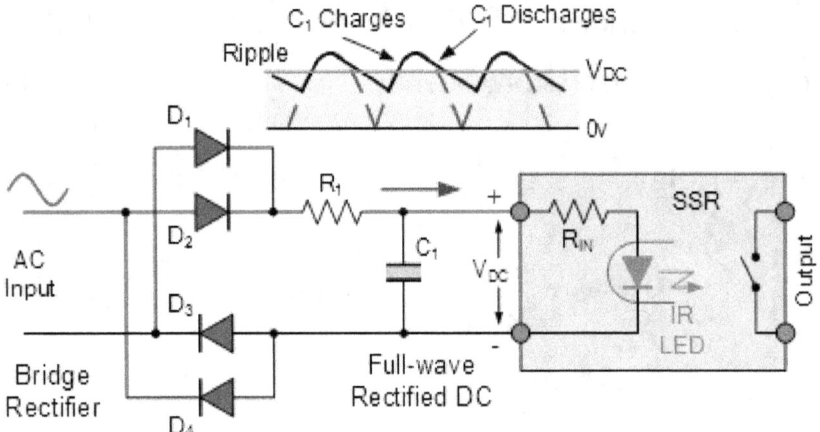

A sinusoidal voltage is converted into full-wave rectified pulses at double the input frequency by bridge rectifiers. The difficulty is that these voltage pulses begin and end at zero volts, which means they will fall below the SSR's input threshold's minimum turn-on voltage requirements, causing the output to turn "on" and "off" every half cycle.

We can smooth out the rectified ripples by putting a smoothing capacitor (C1) on the bridge rectifier's output to overcome the unpredictable output firing. The capacitor's charging and discharging effect will boost the DC component of the rectified signal above the solid state relay's input maximum turn-on voltage. Despite the fact that the sinusoidal voltage waveform is constantly changing, the SSR's input observes a constant DC voltage.

The voltage reducing resistor, R1, and the smoothing capacitor, C1, are selected to match the supply voltage, 120 volts AC or 240 volts AC, as well as the solid state relay's input impedance. However, something in the range of 40k and 10uF would suffice.

A conventional DC solid state relay can therefore be regulated using either an AC or non-polarized DC supply when this bridge rectifier and smoothing capacitor circuit is implemented. Of course, solid state relays with AC input (often 90 to 280 volts AC) are already manufactured and sold.

Output of Solid State Relay

A solid state relay's output switching capability might be AC or DC, depending on the input voltage requirements. Most common solid state relays' output circuits are set up to conduct only one sort of switching action, which is equal to an electromechanical relay's normally-open, single-pole, single-throw (SPST-NO) function.

Power transistors, Darlington's, and MOSFETs are widely used in DC SSRs, whilst a triac or back-to-back thyristors are commonly used in AC SSRs. Because of its high voltage and current characteristics, thyristors are chosen. As demonstrated, a single thyristor can be utilized in a bridge rectifier circuit.

Output Circuit of Solid State Relay

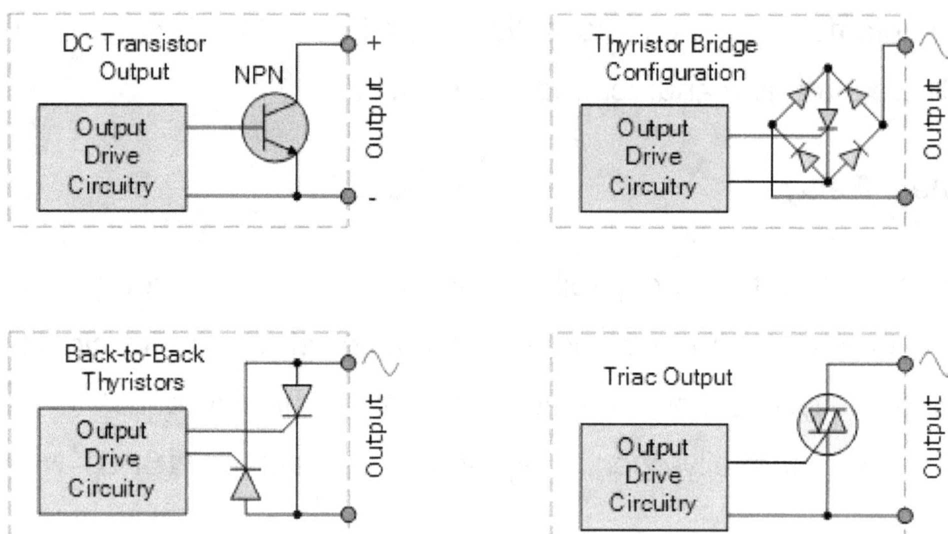

Solid state relays are most widely used to control the switching of an AC load, whether it's for ON/OFF switching, light dimming, motor speed control, or other applications requiring power control. These AC loads can be effectively handled by a low current DC voltage using a solid state relay with a long life and high switching speeds.

One of the most unique features of solid state relays over electromechanical relays is their ability to switch "OFF" AC loads at zero load current, obviating the arcing, electrical noise, and contact bounce that are common with mechanical relays and inductive loads.

This is due to the fact that AC switching solid state relays use SCRs and TRIACs as their output switching devices, which continue to conduct after the input signal has been removed until the AC current flowing through the device falls below its threshold or holding current value.

Then an SSR's output will never turn off in the middle of a sine wave peak. A key benefit of employing a solid state relay is that it decreases electrical noise and back-emf associated with switching inductive loads, which can be realized as arcing by the contacts of an electro-mechanical relay. Have a look at the output waveform diagram of a typical AC solid state relay below.

Output Waveform of Solid State Relay

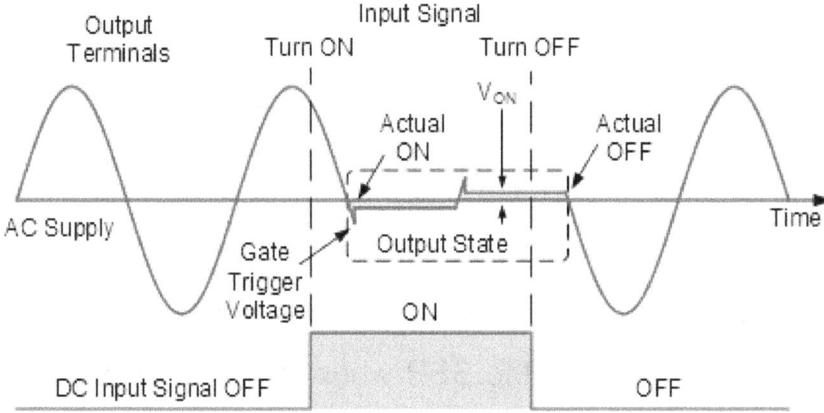

No load current flows through the SSR when no input signal is applied since it is effectively OFF (open-circuited), and the output terminals see the entire AC supply voltage.

With a DC input signal, regardless of which part of the sinusoidal waveform, positive or negative, the cycle is in, the output only turns on when the waveform crosses over the zero point due to the SSR's zero-voltage switching characteristics.

The lowest value required to turn the output thyristors or triacs fully ON is reached as the supply voltage climbs in either a positive or negative direction (usually less than about 15 volts).

The voltage drop across the SSR's output terminals corresponds to the switching device's on-state voltage drop, VT (typically lower than 2 vlots). As a result, any high inrush currents caused by reactive or light loads are significantly decreased.

When the DC input voltage signal is removed, the output does not turn off abruptly because, once triggered into conduction, the thyristor or triac used as the switching device remains ON for the remainder of the half cycle until the load currents fall below the device holding current, at which point it switches OFF. As a result, the large dv/dt back emfs caused by switching inductive loads in the middle of a sine wave are considerably decreased.

The key benefits of the AC solid state relay over the electromechanical relay are its zero crossing function, which switches on the SSR when the AC load voltage is nearly equal to zero volts, suppressing any high inrush currents because the load current will always begin from a point close to 0V, and the thyristor or triac's inherent zero current turn-off characteristic. As a result, there is a maximum turn-off delay of one half cycle (between the elimination of the input signal and the elimination of the load current).

Phase Dimming Solid State Relay

Solid state relays can switch a load in a straight forward zero-crossing manner, but they can also execute far more complex tasks with the help of digital logic circuits, microprocessors, and memories.

Lamp dimming applications, whether in the home or for a play or concert, are another great use for a solid state relay. Non-zero (instant-on) switching solid state relays switch on immediately when the input control signal is applied, as contrast to the zero crossing SSR above, which waits until the AC sine-wave reaches the next zero-crossing point.

This random-fire switching is employed in resistive applications like lamp dimming and applications where the load is only illuminated for a short period of time throughout the AC cycle.

Random Switching Output Waveform

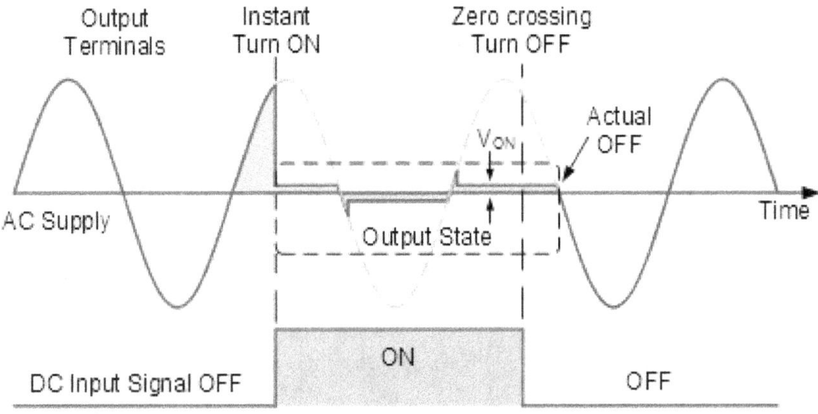

Though this permits for load waveform phase control, the main issue with random turn-on SSRs is that the first load surge current at the time the relay turns on may be significant due to the SSR switching power when the supply voltage is at its peak value (90o).

When the input signal is removed, the load current falls below the holding current of the thyristors or triacs, as indicated. The ON-OFF switching action of a DC SSR is obviously instantaneous. Solid state relays, unlike electro-mechanical relays (EMR), have no moving parts or contacts, making them appropriate for a broad range of ON/OFF switching applications.

However, we may build our own inexpensive and simple solid state relay to regulate an AC load such as a heater, lamp, or solenoid by combining a good opto-isolator with a triac. Since an opto-isolator requires only a modest amount of input/control power to function, the control signal might come from a PIC, Arduino, Raspberry PI, or other similar micro-controller.

Example of Solid State Relay

Let's say we wish to manage a 120V AC, 600 watt heating element with a microcontroller having a digital output port signal of only +5 volts. The MOC 3020 opto-triac isolator might be used for this, but the internal triac can only pass a maximum current (ITSM) of 1 Amps peak at the peak of a 120V AC supply, necessitating the use of a second switching triac.

Let's start with the MOC 3020 opto-input isolator's characteristics (other opto-triacs are available). The forward voltage (VF) drop of the input light emitting diode is 1.2 volts, and the maximum forward current (IF) is 50mA, according to the opto-isolators datasheet.

Up to its maximum value of 50mA, the LED requires roughly 10mA to light reasonably brightly. The microcontroller's digital output port, on the other hand, can only supply 30mA. The needed current is then somewhere between 10 and 30 milliamperes. Hence:

$$R_{MAX} = \frac{V_S - V_F}{10mA} = \frac{5 - 1.2}{0.01} = 380\Omega$$

$$R_{MIN} = \frac{V_S - V_F}{30mA} = \frac{5 - 1.2}{0.03} = 126\Omega$$

As a result, a series current limiting resistor with a value of 126 to 380 can be employed. We will use a preferred resistive value of 240's since the digital output port always switches +5 volts and to reduce power consumption through the opto-couplers LED. This results in a forward current of less than 16mA for the LED.

Any desired resistor value between 150 and 330 would suffice in this case. The heating element has a resistive load of 600 watts. We can get a load current of 5 amperes with a 120V AC source (I = P/V). We'll need a mains switching triac to manage this load current in both half cycles (all four quadrants) of the AC waveform.

The BTA06 is a 600 volt triac with a 6 amp (IT(RMS) rating that can be used for general purpose ON/OFF switching of AC loads, but any equivalent 6 to 8 amp rated triac will suffice. In addition, this switching triac requires just 50mA of gate drive to begin conduction, which is significantly less than the MOC 3020 opto-1 isolator's amp maximum rating.

Assume that the opto-output isolator's triac has turned on at the peak value (90o) of the 120VRMS AC supply voltage. 120 x 1.414 = 170Vpk is the value of this peak voltage. The minimal value of series resistance required is 170/1 = 170's, or 180's to the closest desired number, if the opto-triacs maximum current (ITSM) is 1 ampere peak. On a 120VAC supply, this value of 180s will protect the opto-coupler output triac as well as the gate of the BTA06 triac.

The minimum voltage necessary to generate the requisite 50mA gate drive current forcing the switching triac into conduction is 180 x 50mA = 9.0 volts if the opto-isolator triac switches ON at the zero crossover value (0o) of the 120VRMS AC supply voltage.

When the sinusoidal Gate-to-MT1 voltage exceeds 9 volts, the triac bursts into conduction. As a result, the lowest voltage required after the AC waveform's zero crossover point would be 9 volts peak, and because the power loss in this series gate resistor is so low, a 180 watt rated resistor could be safely employed. Consider the following circuit.

Circuit for AC Solid State Relay

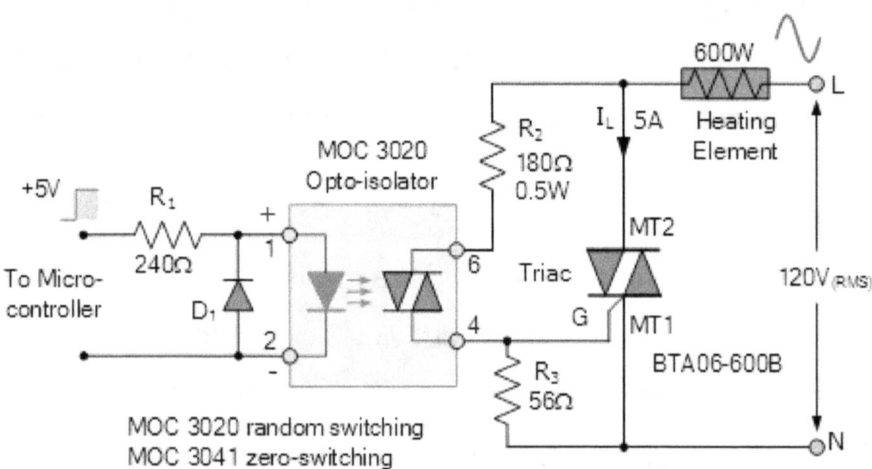

This optocoupler setup is the foundation for a very simple solid state relay application that can control any AC mains driven load, such as lighting and motors. We used the MOC 3020, which is a random switching isolator, in this example. When switching inductive loads, the MOC 3041 opto-triac isolator has the same properties as the MOC 3040, but with built-in zero-crossing detection, letting the load to receive full power without the large inrush currents.

When the triac is turned off, diode D1 prevents harm from reverse connection of the input voltage, while the 56 ohm resistor (R3) shunts any di/dt currents, preventing false triggering. It also connects the gate terminal to MT1, guaranteeing that the triac is entirely turned off. The ON/OFF switching frequency for an AC load should be set to less than 10Hz maximum if utilized with a pulse width modulated, PWM input signal, else the output switching of this solid state relay circuit may not be able to keep up.

CHAPTER-10: SINGLE PHASE RECTIFICATION

Rectification is the process of using solid-state semiconductor devices to connect an AC power supply to a linked DC load. Rectification uses diodes, thyristors, transistors, or converters to convert an oscillating sinusoidal AC voltage source into a constant current DC voltage supply. Half-wave, full-wave, uncontrolled, and fully-controlled rectifiers can transform a single-phase or three-phase supply into a constant DC level.

Rectifiers are one of the most basic components of AC power conversion, with semiconductor diodes performing half-wave or full-wave rectification. Diodes allow alternating currents to flow in one direction while blocking current flow in the other, resulting in a fixed DC voltage level, which makes them perfect for rectification.

Direct current rectified by diodes, on the other hand, is not as pure as direct current obtained from a battery source, and has voltage variations in the form of ripples superimposed on it due to the alternating supply. However, in order to perform single phase rectification, we'll need an AC sinusoidal waveform with a fixed voltage and frequency, as indicated.

AC Sinusoidal Waveform

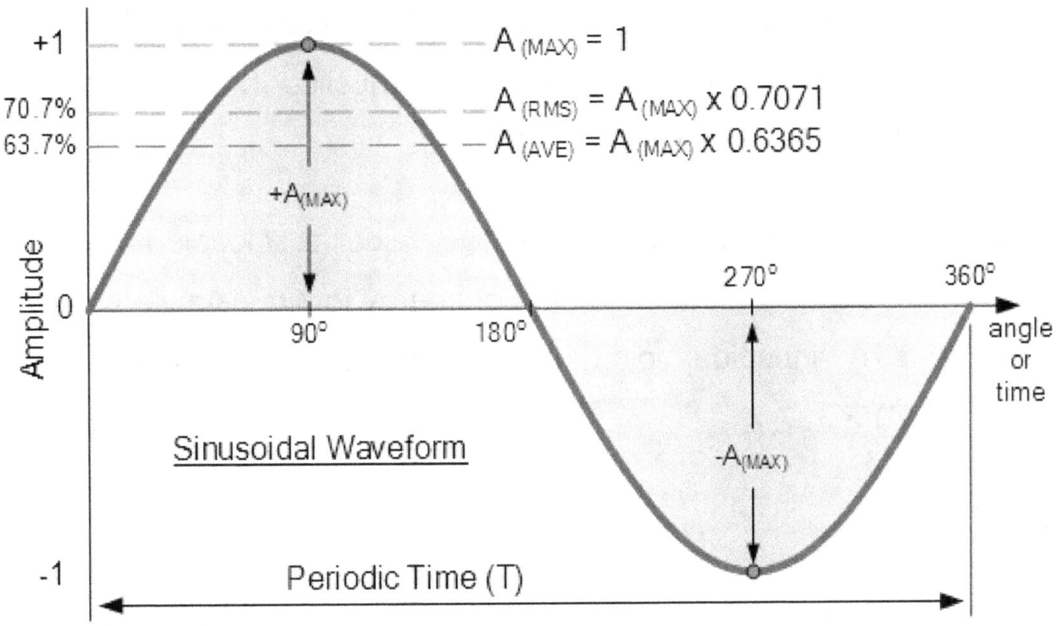

Two numbers are usually linked with AC waveforms. The first value indicates how far the waveform has rotated along the x-axis as the alternator has spun from 0 to 360 degrees. This figure is referred to as the period (T), which is defined as the time it takes for one full cycle of the waveform to finish. Periods can be expressed in degrees, time, or radians. T = 1/ is the formula for the relationship between sine wave periods and frequency.

The second number represents the amplitude of the value along the y-axis, either current or voltage. This number indicates the sine waves highest amplitude from zero to some peak or maximum value (AMAX, VMAX, or IMAX) before returning to zero. There are two maximum or peak values for a sinusoidal waveform, one for the positive half-cycle and one for the negative half-cycle.

But, in addition to these two values, we're interested in two additional for rectification purposes. The average value of the sinusoidal waveform is one, and the RMS value is the other.

The average value of a waveform is 0.6365*VP, which is calculated by adding the instantaneous voltage (or current) values throughout one half-cycle. The average value of a symmetrical sine wave throughout one complete cycle will be zero because the average positive half-wave is cancelled by the opposing average negative half-wave. As a result, +1 + (-1) = 0.

A sinusoid's RMS, root mean squared, or effective value (a sinusoid is another name for a sine wave) sends the same amount of energy to a resistance as a DC supply of same value. A sinusoidal voltage (or current) has a root mean square (rms) value of 0.7071*VP.

Single Phase Rectifier

Solid state devices are used as the principal AC-to-DC converting device in all single phase rectifiers. Since the output is highly influenced by the reactance of the connected load, single phase uncontrolled half-wave rectifiers are the simplest and arguably the most extensively used rectification circuit at small power levels.

Semiconductor diodes are the most widely used device in uncontrolled rectifier circuits, and they are organized in such a way to generate either a half-wave or a full-wave rectifier circuit. The advantage of employing diodes as a rectification device is that they are unidirectional by design, with a one-way pn-junction built in.

By eliminating one-half of the supply, this pn-junction turns a bi-directional alternating supply into a one-way unidirectional current. When the diode is forward-biased, it can pass the positive half of the AC waveform while eliminating the negative half when the diode is reverse-biased.

By removing the positive half of the waveform and passing the negative half, the opposite is also true. In either case, a single diode rectifier's output is only half of the 360o waveform as depicted.

Half-wave Rectification

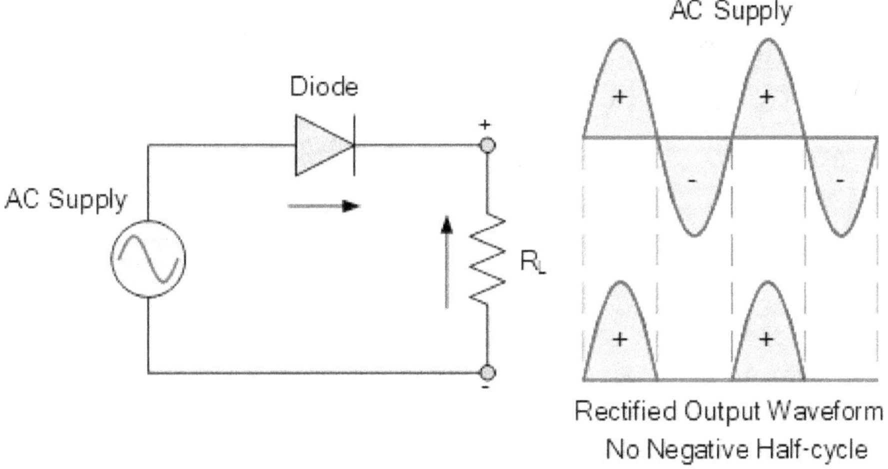

The above single-phase half-wave rectifier design passes the positive half of the AC supply waveform while removing the negative half. We can pass negative parts of the AC waveform and delete positive halves by reversing the orientation of the diode. As a result, the output will consist of a succession of positive or negative pulses.

For half of each cycle, no voltage or current is provided to the associated load, R_L. Since it works for only one-half of the input cycle, the voltage across the load resistance, R_L, comprises of only half waveforms, either positive or negative.

We observed that the diode only permits current to flow in one direction, producing an output consisting of half-cycles. This pulsating output waveform not only varies ON and OFF with each cycle, but it is only present 50% of the time, and with a purely resistive load, this high voltage and current ripple content is at its peak.

Due to the obvious pulsating DC, the corresponding DC value dropped across the load resistor, R_L, is only one-half of the sinusoidal waveform's value. Because the highest value of the sine function of the waveform is 1 (sin(90o), the average or mean DC value obtained over one-half of a sinusoid is defined as: 0.637 x maximum amplitude value. A_{AVE} equals $0.637*A_{MAX}$ during the positive half-cycle. However, since the reverse biased diode eliminates the negative half-cycles, the average value of the waveform during this negative half-cycle is zero, as indicated.

Sinusoids Average Value

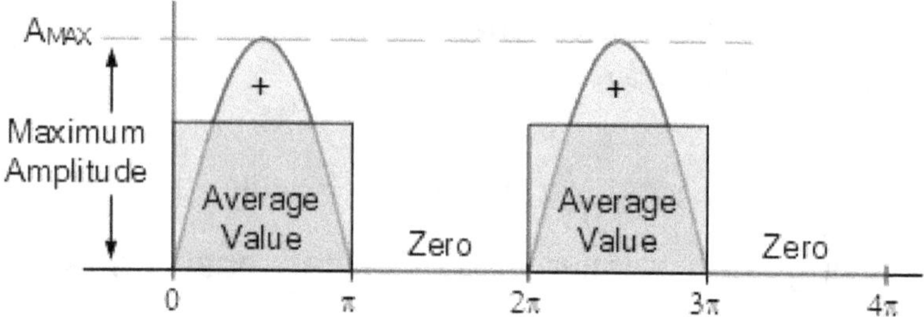

In a half-wave rectifier, the average value is $0.637*A_{MAX}$ 50% of the time, and zero 50% of the time. The average or DC value equivalent across the load resistance, R_L, if the maximum amplitude is 1, will be:

$$A_{AVE} = \frac{0.637}{2} \times A_{MAX} = \frac{A_{MAX}}{\pi} = 0.318 A_{MAX}$$

As a result, for a half-wave rectifier with pulsating DC, the relating expressions for the average value of voltage or current are:

$V_{AVE} = 0.318*V_{MAX}$

$I_{AVE} = 0.318*I_{MAX}$

Although the input waveform's maximum value, A_{MAX}, is the equivalent DC output value of a single phase half-wave rectifier, we could also use its RMS, or "root mean squared" value. To calculate the average voltage for a half-wave rectifier, multiply the RMS value by 0.9 (form factor) and divide by 2, or multiply by 0.45, yielding:

$$V_{AVE} = 0.45*V_{RMS}$$

$$I_{AVE} = 0.45*I_{RMS}$$

Depending on the direction of the diodes, a half-wave rectifier circuit converts either the positive or negative halves of an AC waveform into a pulsed DC output with an equivalent DC value of $0.318*A_{MAX}$ or $0.45*A_{RMS}$, as shown.

Average Voltage of Half-Wave Rectifier

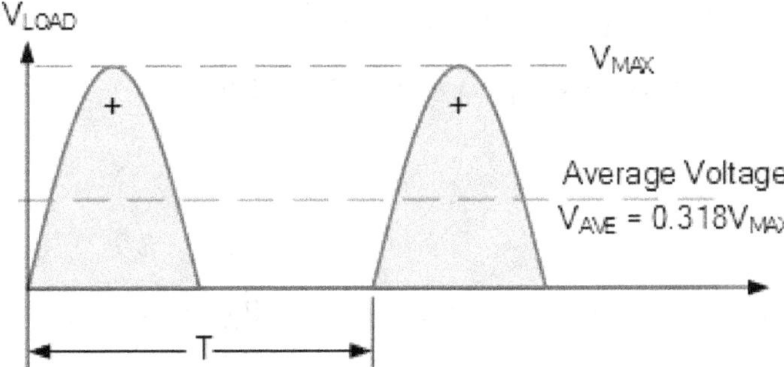

Example of Rectification

A 50V RMS 50Hz AC supply is fed through a single phase half-wave rectifier. If a resistive load of 150 ohms is supplied by the rectifier. Calculate the load current and power dissipated by the load, as well as the equivalent DC voltage developed across it. Assume that you have an ideal diode. To begin, we must convert 50 volts RMS into peak or maximum voltage equivalents.

a) Maximum Voltage Amplitude, V_M

$$V_M = 1.414*V_{RMS} = 1.414*50 = 70.7 \text{ volts}$$

b) Equivalent DC Voltage, V_{DC}

$V_{DC} = 0.318*V_M = 0.318*70.7 = 22.5$ volts

c) Load Current, I_L

$I_L = V_{DC} \div R_L = 22.5/150 = 0.15A$ or $150mA$

d) Power Dissipated by the Load, P_L

$P_L = V*I$ or $I^2*R_L = 22.5*0.15 = 3.375W \cong 3.4W$

Due to the forward biased 0.7 volt voltage drop across the rectifying diode, V_{DC} would be slightly lower in practice. One of the major drawbacks of a single-phase half-wave rectifier is that it produces no output during half of the available input sinusoidal waveform, resulting in a low average value, as we have seen. Using more diodes to create a full-wave rectifier is one way to overcome this.

Full-wave Rectification

The full-wave rectifier, as opposed to the previous half-wave rectifier, uses both halves of the input sinusoidal waveform to produce a unidirectional output. This is due to the fact that the full-wave rectifier is made up of two half-wave rectifiers that are connected together to feed the load.

The single phase full-wave rectifier accomplishes this by using four diodes in a bridge configuration to pass the positive half of the sine wave while inverting the negative half to produce a pulsating DC output.

Despite the fact that the rectifier's voltage and current output is pulsating, it does not reverse direction using the entire input waveform, resulting in full-wave rectification.

Single Phase Full-wave Bridge Rectifier

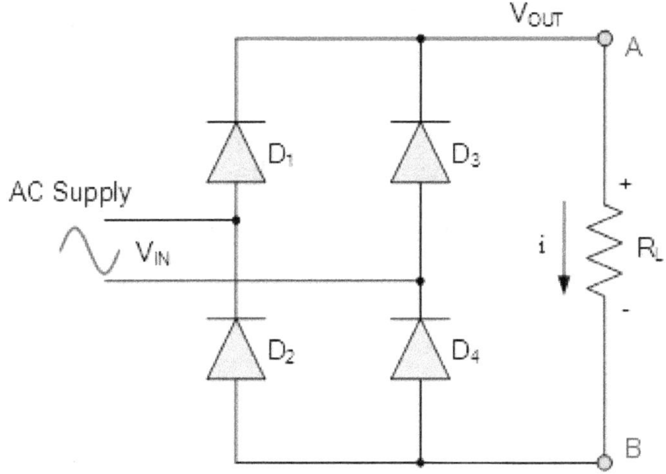

Since two of the four diodes are forward biased and the other two are reverse biased, this bridge configuration of diodes provides full-wave rectification. As a result, instead of a single half-wave rectifier, there are two diodes in the conduction path. Due to the two forward voltage drops of the serially connected diodes, there will be a voltage amplitude difference between V_{IN} and V_{OUT}. As before, we'll assume ideal diodes for the sake of simplicity.

Diodes D_1 and D_4 are forward biased during the positive half cycle of V_{IN}, while diodes D_2 and D_3 are reverse biased.

Current flows along the path $D_1 - A - R_L - B - D_4$ and back to the supply for the positive half cycle of the input waveform.

Diodes D_3 and D_2 are forward biased during the negative half cycle of V_{IN}, while D_4 and D_1 are reverse biased.

Current flows through $D_3 - A - R_L - B - D_2$ and back to the supply for the negative half cycle of the input waveform.

Regardless of the polarity of the input waveform, the positive and negative half-cycles produce positive output peaks in both cases, and as a result, the load current, I always flow in the SAME direction through the load, RL between points

A and B. In this way, the source's negative half-cycle becomes a positive half-cycle at load.

Therefore, node A is always more positive than node B, regardless of which set of diodes is conducting. Hence, the load current and voltage are unidirectional or DC, resulting in the output waveform shown below.

Output Waveform of Full-wave Rectifier

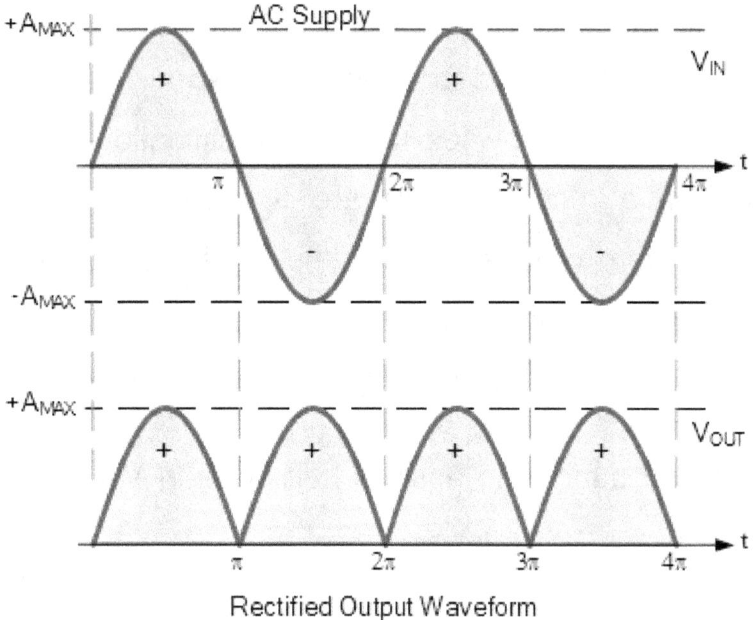

Rectified Output Waveform

The average DC voltage (or current) of this pulsating output waveform is not the same as the input waveform, despite the fact that it uses 100% of the input waveform. The average or mean DC value taken over one-half of a sinusoid is defined as: 0.637 x maximum amplitude value, as we learned earlier. Full-wave rectifiers, on the other hand, have two positive half-cycles per input waveform, resulting in a different average value as shown.

Average Value of Full-wave Rectifier

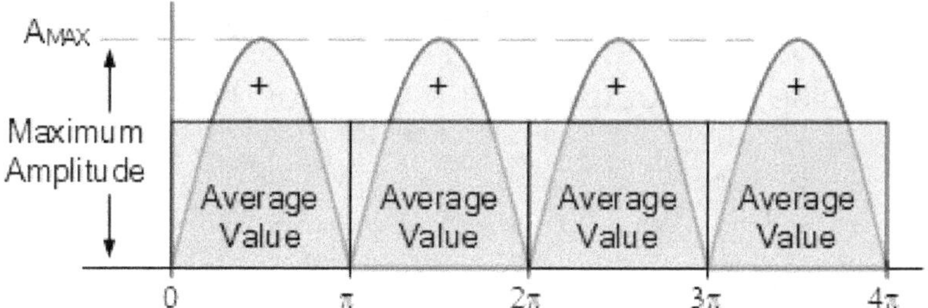

For a full-wave rectifier, the average value for each positive peak is 0.637*A_{MAX}. There are two lots of average value summed together because there are two peaks per input waveform. As a result, a full-wave rectifier's DC output voltage is twice that of the previous half-wave rectifier. The average or DC value equivalent across the load resistance, R_L, if the maximum amplitude is 1, is:

$$A_{AVE} = \frac{2 \times A_{MAX}}{\pi} = \frac{2}{\pi} A_{MAX} = 0.637 A_{MAX}$$

Therefore, the resultant expressions for the average value of voltage or current for a full-wave rectifier is specified as:

$V_{AVE} = 0.637 * V_{MAX}$

$I_{AVE} = 0.637 * I_{MAX}$

The maximum value, A_{MAX}, is the same as before, but we can also use its RMS, or root mean squared value, to find the equivalent DC output value of a single phase full-wave rectifier. To calculate the average voltage for a full-wave rectifier, multiply the RMS value by 0.9, resulting in:

$V_{AVE} = 0.9 * V_{RMS}$

$I_{AVE} = 0.9 * I_{RMS}$

A full-wave rectifier circuit, as shown, converts BOTH the positive and negative halves of an AC waveform into a pulsed DC output with a value of $0.637*A_{MAX}$ or $0.9*A_{RMS}$.

Average Voltage of Full-wave Rectifier

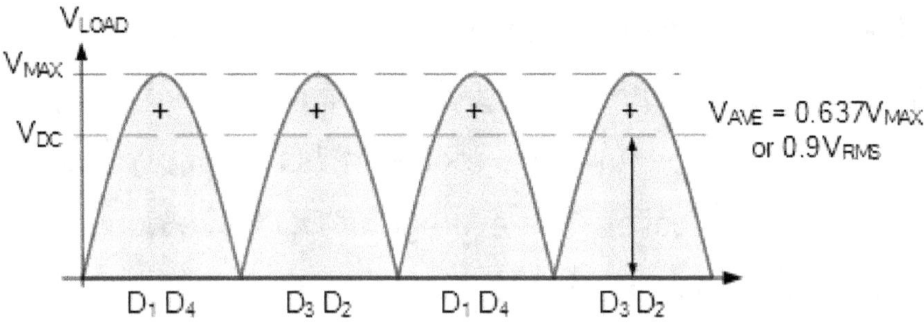

Example of Rectification

A single phase full-wave bridge rectifier circuit using four diodes is required to supply a purely resistive load of 1k at 220 volts DC. Calculate the RMS value of the required input voltage, the total load current drawn from the supply, the load current passed through each diode, and the load's total power dissipated. Assume that the diode has ideal characteristics.

a) Rectifier Supply Voltage, V_{RMS}

$V_{DC} = 0.9*V_{RMS}$ therefore: $V_{RMS} = V_{DC} \div 0.9 = 220/0.9 = 244.4\ V_{RMS}$

b) Load Current, I_L

$I_L = V_{DC} \div R_L = 220/1000 = 0.22A$ or $220mA$

c) Load Current Passed by Each Diode, I_D

The load current is supplied by two diodes per cycle, thus:

$I_D = I_L \div 2 = 0.22/2 = 0.11A$ or 110mA

d) Power Dissipated by the Load, P_L

$P_L = V*I$ or $I^2*R_L = 220*0.22 = 48.4W$

Full-wave Half-Controlled Bridge Rectifier

Full-wave rectification has several advantages over half-wave rectification, including a more consistent output voltage, a higher average output voltage, doubling the input frequency, and requiring a smaller capacitance value smoothing capacitor if one is required.

However, we can improve the design of the bridge rectifier by replacing the diodes with thyristors. We can make a phase-controlled AC-to-DC rectifier by replacing the diodes in a single phase bridge rectifier with thyristors and converting the constant AC supply voltage into a controlled DC output voltage. In variable voltage power supplies and motor control, phase controlled rectifiers, either half-controlled or fully controlled, have a wide range of applications.

The single phase bridge rectifier is a "uncontrolled rectifier" in the sense that the applied input voltage is passed directly to the output terminals, resulting in a fixed average DC equivalent value. It is required to replace two of the diodes with thyristors (SCRs) for converting an uncontrolled bridge rectifier into a single phase half-controlled rectifier circuit.

Half-controlled Bridge Rectifier

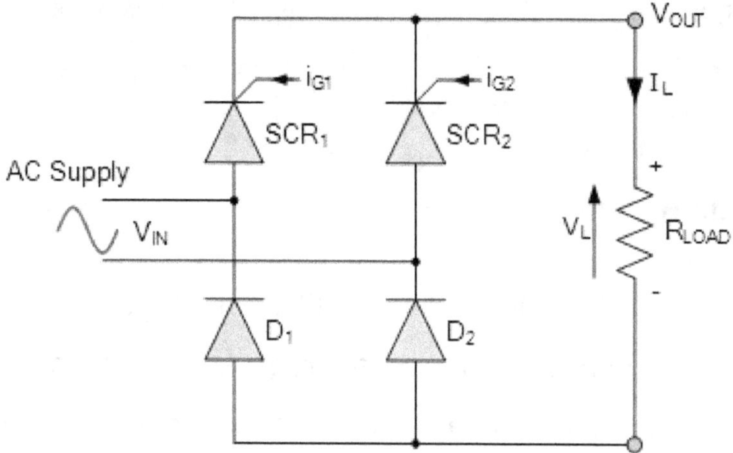

Two thyristors and two diodes control the average DC load voltage in the half-controlled rectifier configuration. When the Anode, (A), is more positive than the Cathode, (K), and a firing pulse is applied to the Gate, (G) terminal, the thyristor will conduct ("ON" state). Else it remains inactive.

Once turned "ON," a thyristor is only turned "OFF" when the gate signal is eliminated and the anode current falls below the thyristor's holding current, IH because the AC supply voltage reverse biases it.

We can control when the thyristor starts to conduct current and thus control the average output voltage by delaying the firing pulse applied to the thyristors gate terminal for a controlled period of time, or angle (α), after the AC supply voltage has passed the zero-voltage crossing of the anode-to-cathode voltage.

Half-controlled Bridge Rectifier

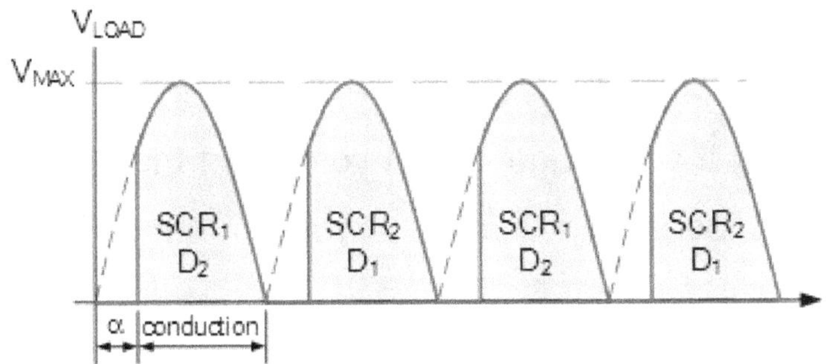

In the positive half cycle of the input waveform, current flows along the path of: SCR_1 and D_2, and return to the supply. In the negative half cycle of V_{IN}, conduction is through SCR_2 and D_1 and return to the supply.

In this case, one thyristor from the top group (SCR_1 or SCR_2) and its resultant diode from the bottom group (D_2 or D_1) must conduct together for any load current to flow.

In this way, the average output voltage, V_{AVE} is reliant on on the firing angle α for the two thyristors included in the half-controlled rectifier as the two diodes are uncontrolled and pass current whenever forward biased. Hence, for any gate firing angle, α, the average output voltage is specified by:

Average Output Voltage of Half-Controlled Rectifier

$$V_{AVE} = \frac{V_{MAX}}{\pi} \times (1 + \cos\alpha)$$

$$\therefore I_{AVE} = \frac{V_{AVE}}{R_L}$$

In this scenario, the maximum average output voltage arises when α = 1 but is still only 0.637*V_{MAX} the same as for the single phase uncontrolled bridge rectifier.

We can take this concept of controlling the average output voltage of the bridge one step further by replacing all four diodes with thyristors providing us a *Fully-controlled Bridge Rectifier* circuit.

Fully-controlled Bridge Rectifier

AC-to-DC converters are more commonly known as single phase fully-controlled bridge rectifiers. Fully-controlled bridge converters are commonly used in the speed control of DC machines, and can be easily made by replacing all four diodes in a bridge rectifier with thyristors, as displayed.

Fully-controlled Bridge Rectifier

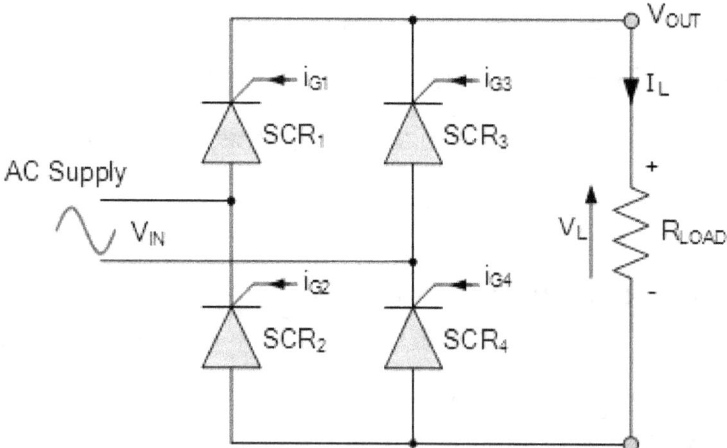

The average DC load voltage is controlled by two thyristors per half-cycle in the fully-controlled rectifier configuration. During the positive half-cycle, thyristors SCR_1 and SCR_4 are fired as a pair, while thyristors SCR_3 and SCR_4 are also fired as a pair during the negative half-cycle. After SCR_1 and SCR_4, it's 180°.

The four thyristors are then switched as alternate pairs to maintain the average or equivalent DC output voltage during continuous conduction mode of operation. The output voltage can be fully controlled, just like the half-controlled rectifier, by varying the thyristors firing delay angle (α). As a result, the average DC voltage from a single phase fully-controlled rectifier in continuous conduction mode can be expressed as follows:

Fully-controlled Rectifier Average Output Voltage

$$V_{AVE} = \frac{V_{MAX}}{\pi} \times \cos(\alpha)$$

$$\therefore I_{AVE} = \frac{V_{AVE}}{R_L}$$

By varying the firing angle α from to π to 0respectively, the average output voltage varies from V_{MAX}/π to $-V_{MAX}/\pi$.

As a result, when the angle is less than 90° (α < 90°), the average DC voltage is positive, and when the angle is greater than 90° (α > 90°), the average DC voltage is negative. Power is transferred from the DC load to the AC supply in this manner.

We observed that in this single phase rectification tutorial, single phase rectifiers can convert AC voltage to DC voltage in a variety of ways, from uncontrolled single diode half-wave rectifiers to fully-controlled full-wave bridge rectifiers using four thyristors. Since only one diode is required, the half-wave rectifier has the advantage of simplicity and low cost. It is inefficient, however, because only half of the input signal is used, resulting in a low average output voltage.

Since it uses both half-cycles of the input sine wave, the full-wave rectifier is more efficient than the half-wave rectifier, resulting in a higher average or equivalent DC output voltage. The full-wave bridge circuit has the disadvantage of requiring four diodes.

Phase controlled rectification converts the AC input voltage to a controlled DC output voltage using a combination of diodes and thyristors (SCRs). Half-controlled rectifiers use a combination of thyristors and diodes in their configuration, whereas fully-controlled rectifiers use four thyristors.

Rectification is the process of converting a sinusoidal AC waveform to a steady state DC supply, regardless of how we do it.

CHAPTER-11: THREE PHASE RECTIFICATION

The process of converting a balanced 3-phase power supply into a fixed DC supply using solid state diodes or thyristors is known as 3-phase rectification. Rectification is the process of converting an AC input supply into a fixed DC supply, as we saw in the previous tutorial. One of the most common circuits used to perform this rectification is one based on solid-state semiconductor diodes.

Rectification of alternating voltages is one of the most common practices for diodes, because diodes are inexpensive, small, and robust, allowing us to build a wide range of rectifier circuits with either individually connected diodes or a single integrated bridge rectifier module.

Single phase supplies in homes and offices are typically 120 Vrms or 240 Vrms phase-to-neutral, also known as line-to-neutral (L-N), and nominally of a fixed voltage and frequency, producing an alternating voltage or current in the form of a sinusoidal waveform, abbreviated "AC."

Three-phase rectification, also referred as poly-phase rectification circuits, are similar to single-phase rectifiers; the difference is that we're using three single-phase supplies connected together that were all generated by a single three-phase generator this time.

The benefit is that 3-phase rectification circuits can be used to power a wide range of industrial applications, including motor control and battery charging, that require more power than a single-phase rectifier circuit can provide.

Three-phase supplies take this concept a step further by combining three AC voltages of identical frequency and amplitude, each of which is referred to as a "phase."

These three phases are 120 degrees electrically out of phase with one another, resulting in a phase sequence, or phase rotation, of 360° ÷ 3 = 120°, as shown. Three-phase Waveform

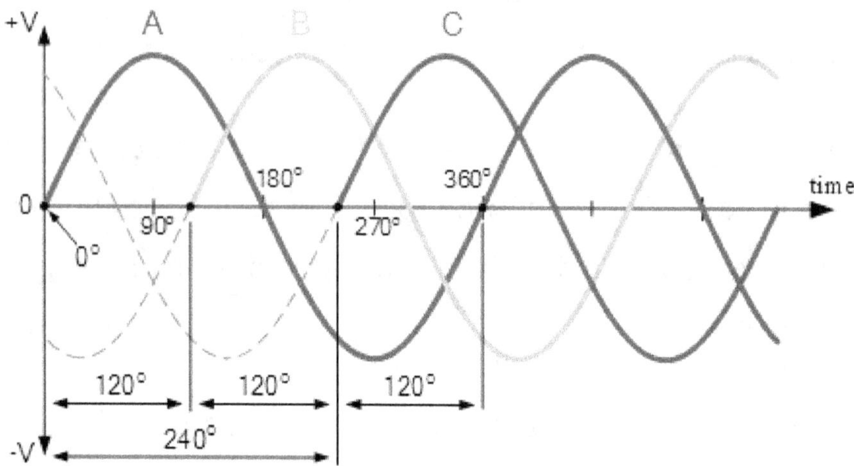

A three-phase alternating current (AC) supply can be used to deliver electrical power directly to balanced loads and rectifiers which is beneficial.

Since a 3-phase supply has a fixed voltage and frequency, it can be used by a rectification circuit to generate a fixed voltage DC power that can then be filtered, resulting in a lower ripple output DC voltage than a single-phase rectifying circuit.

Three-phase Rectification

We can use this multi-phase features to build 3-phase rectifier circuits now that we know a 3-phase supply is just three single-phases combined together.

Three-phase rectification, like single-phase rectification, employs diodes, thyristors, transistors, or converters to construct half-wave, full-wave, uncontrolled, and fully-controlled rectifier circuits that convert a three-phase supply to a constant DC output level.

In most cases, a three-phase rectifier is powered directly from the utility power grid or from a three-phase transformer if the connected load requires a different DC output level.

The most basic three-phase rectifier circuit, like the previous single-phase rectifier, is an uncontrolled half-wave rectifier circuit that uses three semiconductor diodes, one for each phase, as shown.

Half-Wave Three-phase Rectification

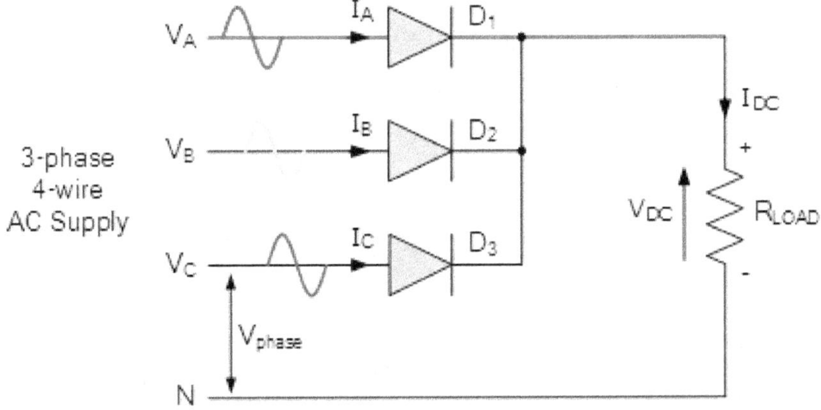

Each diode's anode is connected to one phase of the voltage supply, and the cathodes of all three diodes are connected to the same positive point, resulting in a diode-"OR" type arrangement. The positive (+) terminal of the load is connected to this common point, while the negative (-) terminal is connected to the supply's neutral (N).

Presuming a phase rotation of Red-Yellow-Blue ($V_A - V_B - V_C$) and the red phase (V_A) starts at 0°. The first diode to conduct will be diode 1 (D_1) as it will have a more positive voltage at its anode than diodes D_2 or D_3. Thus diode D_1 conducts for the positive half-cycle of V_A while D_2 and D_3 are in their reverse-biased state. The neutral wire delivers a return path for the load current back to the supply.

120 electrical degrees later, diode 2 (D_2) starts to conduct for the positive half-cycle of V_B (yellow phase). Now its anode becomes more positive than diodes D_1 and D_3 which are both "OFF" because they are reversed-biased. Likewise, 120° later V_C (blue phase) starts to increase turning "ON" diode 3 (D_3) as its anode becomes more positive, thus turning "OFF" diodes D_1 and D_2.

We observed that in three-phase rectification, whichever diode has a higher positive voltage at its anode than the other two diodes will begin to conduct automatically, thereby providing a conduction form of: D_1 D_2 D_3 as shown.

Conduction Waveform of Half-Wave Three-phase Rectifier

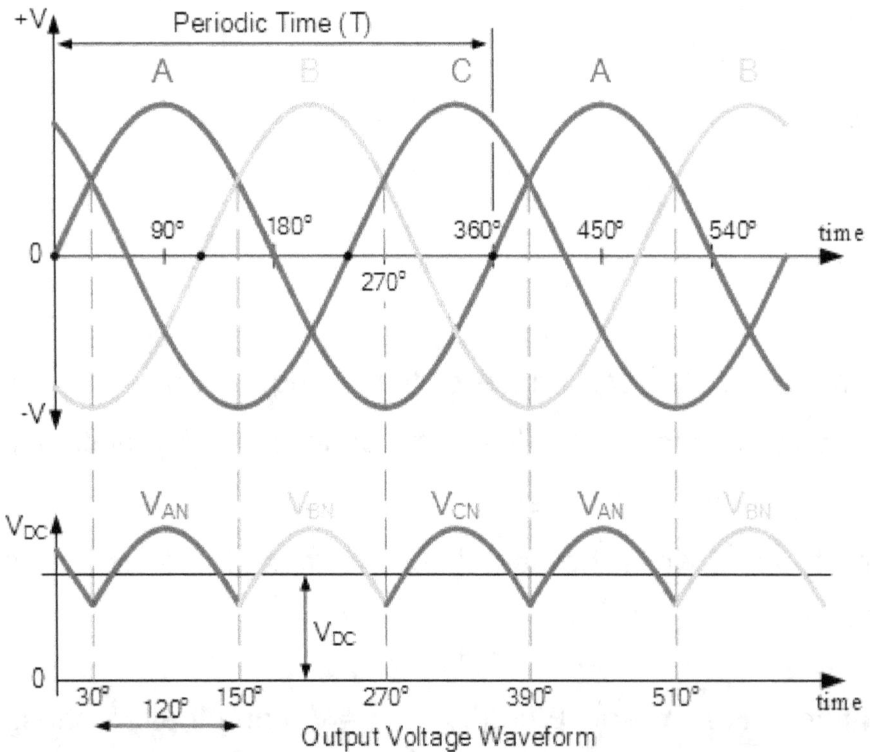

We can observe from the above waveforms for a resistive load that each diode in a half-wave rectifier passes current for one third of each cycle, and the output waveform is three times the AC supply's input frequency.

As a result, each cycle has three voltage peaks, so increasing the number of phases from a single-phase to a three-phase supply improves the rectification of the supply, resulting in a smoother output DC voltage.

In a three-phase half-wave rectifier, the supply voltages V_A V_B and V_C are balanced but with a phase difference of 120° providing:

$V_A = V_P*\sin(\omega t - 0°)$

$V_B = V_P*\sin(\omega t - 120°)$

$V_C = V_P*\sin(\omega t - 240°)$

In this way, the average DC value of the output voltage waveform from a 3-phase half-wave rectifier is specified as:

$$V_{DC} = \frac{3\sqrt{3}}{2\pi}V_P = 0.827*V_{PEAK}$$

Since V_P is equal to $V_{RMS}*1.414$ when the voltage delivers peak voltage, V_{RMS} is equal to $V_P/1.414$, or $0.707*V_P$ when $1/1.414 = 0.707$. The rectifier's average DC output voltage can then be expressed as follows in terms of its root-mean-squared (RMS) phase voltage:

$$V_P = 1.414 \times V_{RMS} \qquad \therefore V_{RMS} = \frac{V_P}{1.414}$$

$$V_{DC} = \frac{3\sqrt{3}}{2\pi} \times V_P = 0.827 \times V_P$$

Convert to V_{RMS}

$$V_{DC} = 0.827 \times 1.414 \times V_{RMS}$$

$$\therefore V_{DC} = \frac{0.827}{0.707} V_{RMS} = 1.17 \times V_{RMS}$$

Example of 3-phase Rectification

Three individual diodes and a 120VAC 3-phase star connected transformer are used to build a half-wave 3-phase rectifier. If a connected load with an impedance of 50Ω must be powered,

a) Determine the average DC voltage delivered to the load.

b) the load current; and c) the average current per diode

Consider that diodes are ideal.

a). The average DC load voltage:

$$V_{DC} = 1.17*Vrms = 1.17*120 = 140.4 \text{ volts}$$

Note that if we were given the peak voltage (V_p) value, then:

V_{DC} would equal $0.827*Vp$ or $0.827*169.68 = 140.4V$.

b). The DC load current:

$$I_L = V_{DC}/R_L = 140.4/50 = 2.81 \text{ amperes}$$

c). The average current per diode:

$$I_D = I_L/3 = 2.81/3 = 0.94 \text{ amperes}$$

A 4-wire supply with three phases plus a neutral (N) connection is required for half-wave 3-phase rectification. Furthermore, with a value of $0.827*V_P$, the average DC output voltage is low. The output ripple frequency is three times that of the input, which accounts for this. However, we can build a three-phase full-wave uncontrolled bridge rectifier that overcomes these drawbacks by adding three additional diodes to the basic rectifier circuit.

Full-wave Three-phase Rectification

The full-wave three-phase uncontrolled bridge rectifier circuit uses six diodes, two per phase, similar to the single-phase bridge rectifier. A three-phase full-wave rectifier is built using two half-wave rectifier circuits. The circuit produces less ripple than the previous half-wave 3-phase rectifier because its frequency is six times that of the input AC waveform. The full-wave rectifier can also be powered by a balanced 3-phase 3-wire delta connected supply because no fourth neutral (N) wire is required. Consider the full-wave 3-phase rectifier circuit shown below.

Full-wave Three-phase Rectification

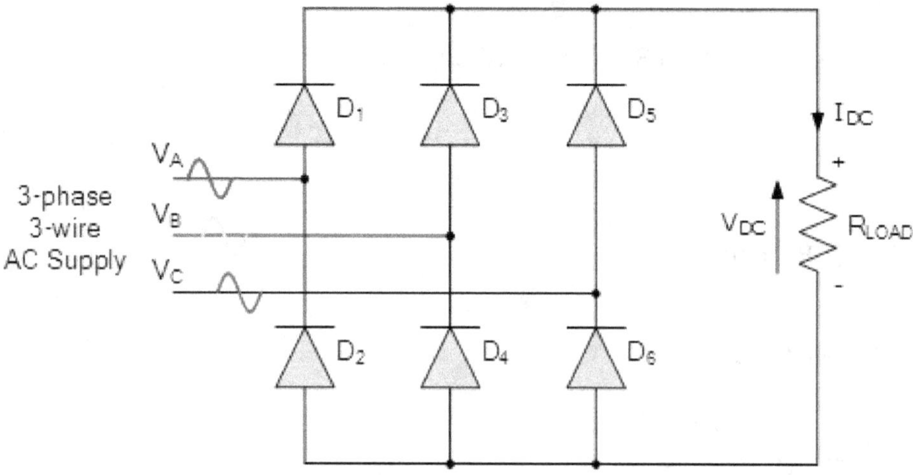

Let's consider a phase rotation of Red-Yellow-Blue ($V_A - V_B - V_C$) and the red phase (V_A) starts at 0°. Each phase connects between a pair of diodes as shown. One diode of the conducting pair powers the positive (+) side of load, while the other diode powers the negative (-) side of load.

Diodes D_1 D_3 D_2 and D_4 form a bridge rectifier network between phases A and B, similarly diodes D_3 D_5 D_4 and D_6 between phases B and C and D_5 D_1 D_6 and D_2 between phases C and A.

Diodes D_1 D_3 and D_5 feed the positive rail. The diode which has a more positive voltage at its anode terminal conducts. Likewise, diodes D_2 D_4 and D_6 feed the negative rail and whichever diode has a more negative voltage at its cathode terminal conducts.

The diodes conduct in matching pairs for three-phase uncontrolled rectification, with each conduction path passing through two diodes in series. With commutation of the circuit occurring every 60°, or six times per cycle, a total of six rectifier diodes are required.

If we start the form of conduction at 30°, this gives us a conduction form for the load current of: D_{1-4} D_{1-6} D_{3-6} D_{3-2} D_{5-2} D_{5-4} and return again to D_{1-4} and D_{1-6} for the next phase sequence as shown.

Full-wave Three-phase Rectifier Conduction Waveform

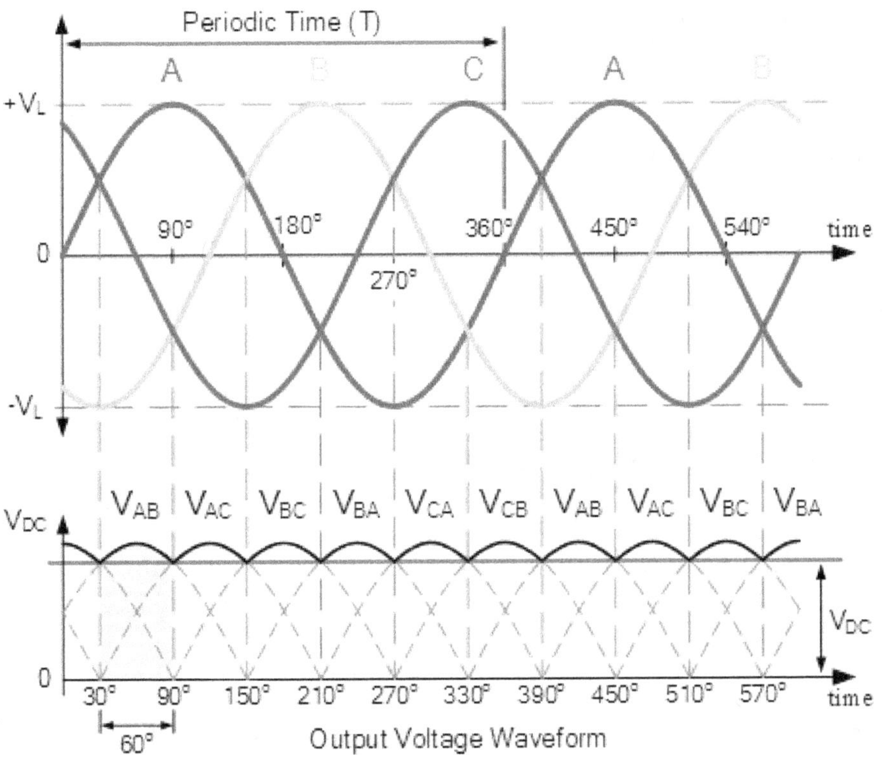

Conduction always occurs in the most positive diode and the corresponding most negative diode in 3-phase power rectifiers. Conduction is passed from diode to diode as the three phases rotate across the rectifier terminals. In each supply cycle, each diode conducts for 120° (one-third), but because it takes two diodes to conduct in pairs, each pair of diodes conducts for only 60° (one-sixth) of a cycle at any given time, as shown above. As a result, we can correctly state that for a 3-phase rectifier fed by "3" transformer secondary, each phase will be separated by 360o/3, necessitating the use of 2*3 diodes.

There is also no common connection between the rectifier's input and output terminals, unlike the previous half-wave rectifier. So, it can be powered by either a star or a delta connected transformer supply. Hence, the average DC value of a 3-phase full-wave rectifier's output voltage waveform is:

$$V_{DC} = \frac{3\sqrt{3}}{\pi} V_S = 1.65 * V_S$$

Here: V_S is equal to ($V_{L(PEAK)} \div \sqrt{3}$) and where $V_{L(PEAK)}$ is the maximum line-to-line voltage ($V_L * 1.414$).

Example of 3-phase Rectification

To feed a 150 resistive load from a 3-phase 127 volt, 60Hz delta connected supply, a 3-phase full-wave bridge rectifier is required.

Calculate: 1. the rectifier's DC output voltage and 2. the load current, ignoring the voltage drops across the diodes.

1. the DC output voltage:

The RMS (Root Mean Squared) line voltage is 127 volts. So, the line-to-line peak voltage ($V_{L-L(PEAK)}$) will be:

$$V_{L(PEAK)} = V_{L(RMS)} \times \sqrt{2} = 127 \times 1.414 = 179.6\,V$$

Since, the supply is 3-phase, the phase to neutral voltage (V_{P-N}) of any phase will be:

$$V_S = V_{L(PEAK)} \div \sqrt{3} = 179.6 \div 1.732 = 103.7\,V$$

$$V_S = \frac{V_{L(RMS)} \times \sqrt{2}}{\sqrt{3}} = 103.7\,V$$

In this way, the average DC output voltage from the 3-phase full-wave rectifier is specified as:

$$V_{DC} = \left[\frac{3\sqrt{3}}{\pi}\right]V_S = 1.654 \times V_S$$

$$\therefore V_{DC} = 1.654 \times 103.7 = 171.5\,V$$

For a given line-to-line RMS voltage value of 127 volts, the average DC output voltage is:

$$V_{DC} = \frac{3\sqrt{2}}{\pi}V_{L(RMS)} = 1.35 \times 127 = 171.5\,V$$

2. Load current of the rectifier

The output from the rectifier is feeding a 150Ω resistive load. Then using Ohms law the load current will be:

$$I_{LOAD} = V_S \div R_L = 171.5 \div 150 = 1.14\ \text{Amps}$$

Uncontrolled 3-phase rectification employs diodes to produce a fixed average output voltage in relation to the input AC voltages. To vary the rectifier's output voltage, we must replace the uncontrolled diodes with thyristors, either some or all of them, to create half-controlled or fully-controlled bridge rectifiers.

Thyristors are three-terminal semiconductor devices that conduct and pass a load current when a suitable trigger pulse is applied to the gate terminal when the Anode–to–Cathode terminal voltage is positive. We can delay the instant in time when the thyristor would naturally switch "ON" if it were a normal diode and the moment it starts to conduct when the trigger pulse is applied by delaying the timing of the trigger pulse (firing angle).

Thus, with a controlled 3-phase rectification using thyristors rather than diodes, we can control the value of the average DC output voltage by controlling the firing angle of the thyristor pairs, and the rectified output voltage becomes a function of the firing angle, α.

As a result, the only difference between the formula above for the average output voltage of a 3-phase bridge rectifier and the formula above for the average output voltage of a 3-phase bridge rectifier is the cosine angle, $\cos(\alpha)$, of the firing or triggering pulse.

As a result, if the firing angle is zero ($\cos(0) = 1$), the controlled rectifier performs similarly to the previous three-phase uncontrolled diode rectifier, with the same average output voltages.

The following is an example of a fully-controlled 3-phase bridge rectifier:

Fully-controlled 3-phase Bridge Rectifier

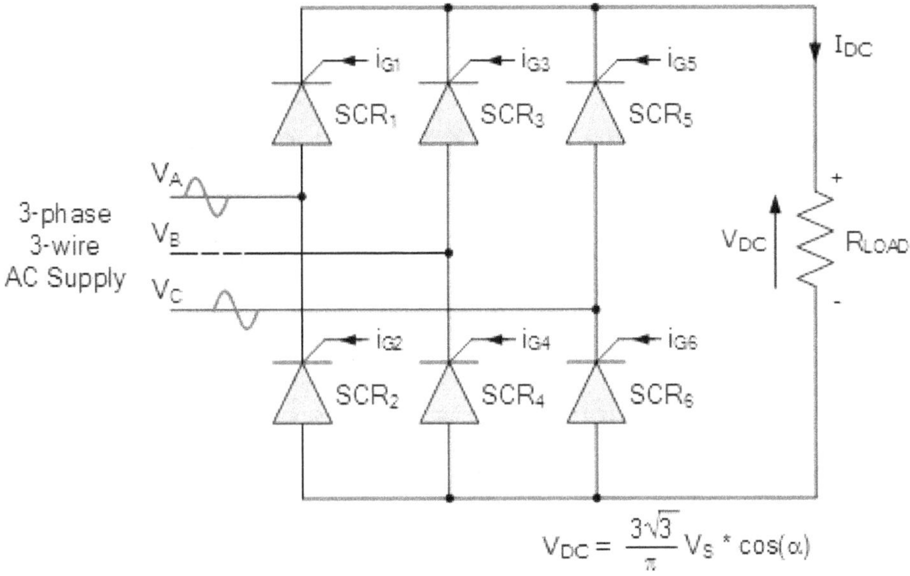

$$V_{DC} = \frac{3\sqrt{3}}{\pi} V_S * \cos(\alpha)$$

Three-phase Rectification in A Brief

Three-phase rectification, as we've seen in this tutorial, is the process of converting a three-phase AC supply into a pulsating DC voltage, as rectification converts a sinusoidal voltage and frequency into a fixed voltage DC power.

As a result of power rectification, an alternating supply is converted to a unidirectional supply. However, we've seen that 3-phase half-wave uncontrolled rectifiers, which use one diode per phase, require a fourth neutral (N) wire to close the circuit from load to source. Three mains lines are required for a three-phase full-wave bridge rectifier with two diodes per phase, such as those provided by a delta connected supply.

In comparison to a half-wave bridge rectifier, another advantage of a full-wave bridge rectifier is that the load current is well balanced across the bridge, improving efficiency (the ratio of output DC power to input DC power supplied) and reducing ripple content, both in amplitude and frequency.

It is possible to obtain a higher average DC output voltage with less ripple amplitude by increasing the number of phases and diodes within the bridge configuration. For example, in 6-phase rectification, each diode would conduct for only one-sixth of a cycle.

Multi-phase rectifiers also produce a higher ripple frequency, which results in less capacitive filtering and a smoother output voltage. In this way, uncontrolled rectifiers with 6, 12, 15, and even 24-phases can be designed to improve the ripple factor for a variety of applications.